GLOBAL PHILOSOPHY

What Philosophy Ought to Be

NICHOLAS MAXWELL

SOCIETAS
essays in political
& cultural criticism

imprint-academic.com

Copyright © Nicholas Maxwell, 2014

The moral rights of the author have been asserted.
No part of this publication may be reproduced in any form
without permission, except for the quotation of brief passages
in criticism and discussion.

Published in the UK by
Imprint Academic, PO Box 200, Exeter EX5 5YX, UK

Distributed in the USA by
Ingram Book Company,
One Ingram Blvd., La Vergne, TN 37086, USA

ISBN 9781845407674

A CIP catalogue record for this book is available from the
British Library and US Library of Congress

Other titles by Nicholas Maxwell:

What's Wrong With Science?
From Knowledge to Wisdom
The Comprehensibility of the Universe
The Human World in the Physical Universe
Is Science Neurotic?
Cutting God in Half – and Putting the Pieces Together Again
How Universities Can Help Create a Wiser World: The Urgent Need for an Academic Revolution
With R. Barnett, ed., *Wisdom in the University*
L. McHenry, ed., *Science and the Pursuit of Wisdom: Studies in the Philosophy of Nicholas Maxwell*

Contents

Preface: Learning, Global Problems, and Play — vii

Chapter 1: Philosophy Seminars for Five-Year-Olds — 1

Chapter 2: What Philosophy Ought to Be — 11

Chapter 3: How Can Our Human World Exist and Best Flourish Embedded in the Physical Universe? A Letter to an Applicant to a New Liberal Studies Course — 47

Chapter 4: What's Wrong with Science and Technology Studies? What Needs to Be Done to Put It Right? — 65

Chapter 5: Arguing for Wisdom in the University: An Intellectual Autobiography — 108

Acknowledgements — 177

References — 178

Index — 187

Preface

Learning, Global Problems, and Play

These essays are about education, learning, rational inquiry, philosophy, science studies, problem solving, academic inquiry, global problems, wisdom and, above all, the urgent need for an academic revolution.

Despite this range and diversity of topics, there is a common underlying theme. Education ought to be devoted, much more than it is, to the exploration of real-life, open problems; it ought not to be restricted to learning up solutions to already solved problems — especially if nothing is said about the problems that provoked the solutions in the first place. There should be much more emphasis on learning how to engage in cooperatively rational exploration of problems: even five-year-olds could begin to learn how to do this. A central task of philosophy ought to be to keep alive awareness of our unsolved fundamental problems — especially our most fundamental problem of all, encompassing all others: How can our human world — and the world of sentient life more generally — imbued with the experiential, consciousness, free will, meaning, and value, exist and best flourish embedded as it is in the physical universe? This is both our fundamental intellectual problem and our fundamental problem of living.

As far as the latter is concerned, we are at present heading towards disaster — as our immense, unsolved global problems tell us: population growth, destruction of natural habitats and rapid extinction of species, vast inequalities of wealth and power around the world, pollution of earth, sea, and air, our proclivity for war, and above all global warming. If we are to resolve our conflicts and global problems more intelligently, effectively, and humanely than we have managed to do so far, then we have to learn how to do it. That, in turn, requires that our institutions of learning, our universities and schools, are rationally designed and devoted to the task. At present they are not. That is the crisis behind all the others. From the past we have inherited the idea that the basic intellectual aim of inquiry ought to be to acquire knowledge. First, knowledge is to be acquired; then, secondarily, it can be applied to help solve social problems. But this is dangerously and damagingly irrational, and it is this irrationality that is, in part, responsible for the genesis of our current global problems, and our current incapacity to solve them. As a matter of supreme urgency, we need to transform academia so that it becomes rationally devoted to helping humanity learn how to make progress towards as good and wise a world as possible. This would involve putting problems of living—including global problems—at the heart of academia, problems of knowledge and technological know-how emerging out of, and feeding back into, the central task to help people tackle problems of living in increasingly cooperatively rational ways. Almost every department and aspect of academia needs to change. We need a new kind of academic inquiry devoted not just to knowledge but rather to wisdom—wisdom being the capacity to realize what is of value in life for oneself and others, wisdom including knowledge and technological know-how, but much else besides.

So, this is what these essays seek to provoke: a concerted effort to transform our institutions of learning so that they become rationally and effectively devoted to helping us learn how to create a wiser world.

With these essays before me, I can see that there is one crucial element of learning about which they say nothing—or nothing explicit. The vital role of play in learning. All mammals—or at any rate almost all mammals—learn by means of play. Cats, tigers, foxes, and other predators learn to hunt by means of endless mock fights when kittens and cubs. Deer, sheep, and antelope learn to escape by means of playful leaps and bounds when young. We are mammals too. Almost certainly, we learnt how to be adult human beings by means of play during the millions of years we evolved into *homo sapiens* living in hunting and gathering tribes. Children today, out of school, learn by means of play. Learning by means of play is almost certainly fundamental to our make-up. Education needs to exploit it. Schools and universities need to become places of play. Successful problem solving is often likely to be playful in character. The youthful Einstein called doing physics "getting up to mischief".

But our most serious problems of living are so grim, so imbued with suffering, wasted lives, and unnecessary death, that the idea of approaching them in a playful spirit seems sacrilegious. We need to keep alive tackling of intellectual problems so that playful capacities can be exercised—if for no other reason (and other reasons there are, of course, aplenty). There are two really worthy impulses behind all rational inquiry: delight and compassion.

Chapter One
Philosophy Seminars for Five-Year-Olds

For Harry

All of us, I believe, are extraordinarily active and creative intellectually when we are very young. Somehow, in the first few years of life, we acquire an identity, a consciousness of self; we discover, or create, a whole view of the world, a cosmology; and we learn to understand speech, and to speak ourselves. And we achieve all this without any formal education whatsoever. Compared with these mighty intellectual achievements of our childhood, the heights of adult artistic and scientific achievement all but pale into insignificance. It is reasonable to suppose that there is a biological, a neurological, basis for our extraordinary capacity to learn when we are very young. It probably has to do with the fact that our brains are still growing during the first few years of life. It is striking that there are things that can only be learnt during this time. If we have not had the opportunity to learn to speak by the age of twelve, we will never really learn to speak. Lightning calculators all begin to acquire their extraordinary arithmetical skills when very young. Some things, it seems, become too difficult for us to learn as we grow older. In our early childhood we are forced, by our situation, to be creative philosophers and metaphysicians, preoccupied by fundamental issues. One has only to think of the endless questioning of young children to appreciate something of their insatiable hunger to know, to understand.

The tragedy is that formal education so rarely helps us to recognize and to develop our early profound intellectual experiences and achievements. Instead of encouraging our instinctive curiosity to develop into adulthood, all too often education unintentionally stifles and crushes it out of existence.

Academic inquiry *ought* to be the outcome of all our efforts to discover what is of value in existence and to share our discoveries with others. At its most important and fundamental, inquiry is the thinking we engage in as we live, as we strive to realize what is of value to us in our life. All of us ought both to contribute to and to learn from interpersonal public inquiry. This two-way traffic of teaching and learning ought to start at the outset, when we first attend school. Young children, at school, need to be encouraged to tell each other about their discoveries, their experiences, their thoughts and problems. The teacher needs to encourage both speaking and listening. Such a class or seminar, devoted to the cooperative, imaginative, and rational exploration of problems encountered in life, ought to form a standard—even a central and fundamental—part of all education, science, and scholarship, from primary school to university.

If this were the case, then we might all discover how to use science and scholarship so as to develop our own thinking—and living. Telling others of our problems and ideas—and listening to others tell of theirs—would help us to discover and to value our own thinking. It is all too easy to dismiss our most serious and original thinking—those moments of bafflement, surmise, and wonder—as mere wordless feeling, irredeemably private, signifying little. This is especially the case in childhood. Unarticulated, our thinking is liable to become neglected, stagnant, forgotten. If it is to flourish it is vital that we develop and constantly practise the difficult art of putting what we feel and think into public words. An education that gave an intellectually fundamental role to the development of this art would not only stimulate the growth of personal thinking, it would also enable us to discover vital interconnections between our personal thinking and public scientific and scholarly thought.

Academic education would be not an imposition but an invitation to participate from the outset.

I do not want to exaggerate. Education of this person-centred, participatory kind already exists, to some extent, in both the arts and the sciences. Teachers of literature, drama, and the other arts appreciate that art serves, as it were, a double purpose. As we enhance our understanding and appreciation of literature, so too, incidentally, we may enhance our understanding of ourselves and of others. By exploring, in novels and plays, imaginary people living imaginary lives, we can achieve a freedom to explore aspects of ourselves without the embarrassment or torture of public self-exposure. Furthermore, in order to improve our understanding of literature it is important that we try our hand at writing, which can enhance our powers of self-expression and our self-understanding. Analogous remarks can be made about drama, art, music, dance. And again, in science education at its best, it is appreciated that it is not just scientific results that need to be taught, but also, and perhaps most fundamentally, scientific *problems*. It has long been appreciated that in order to *understand* science it is essential to *do* it.

What is missing in all this is an appreciation of the central and unifying role of *philosophy* in all of education—philosophy pursued as the cooperative, imaginative, and rational exploration of fundamental problems of living. Philosophy pursued in this way would effortlessly bridge the gulf between science and art, science and the humanities. All other parts of the curriculum—the physical and biological sciences, mathematics, geography, history, politics, literature, theatre, religion, etc.—could quite naturally and understandably emerge out of, and feed back into, the central, unifying enterprise of philosophy pursued as the open, rational exploration of fundamental problems. The very problem of how to unify all the diverse aspects of the world into a coherent, understandable whole could itself be recognized and discussed. The world we live in is a more or less interconnected whole: it is not experienced as being split up into physics, chemistry, biology, history, literature, religion, and so on. Setting out to improve children's knowledge and understanding of the world in specialized, dissociated fragments, without any indication as to how the fragments fit together or, worse, without even an indication of the existence of the problem, is in itself an appallingly anti-rational and alienating thing to do. It amounts to the imposition of a sort of intellectual schizophrenia. It sets up a barrier between *personal* thinking and departmentalized *academic* thought, resulting in mutual distrust rather than mutual enhancement between these two kinds of thought. In important respects, academic learning cannot promote—it can only sabotage—coherent, rational thought about problems of living in this one, real, interconnected world.

Failure to teach philosophy to five-year-olds, as a central, unifying part of the curriculum, is the result of mistaken assumptions about both children and philosophy.

Philosophy, it is assumed, is too difficult and esoteric a subject to be taught to five-year-olds. Only adults can come to grips with such an advanced discipline. In fact it is, if anything, the other way round. Above all, it is young children who are compelled, by their situation, to be highly active and creative philosophers, daily concerned with fundamental problems about the nature of life and the world. Most adults have long ago settled in their minds, in one way or another, fundamental questions about the nature of life and the world. It is particular, detailed, and specialized problems that preoccupy adult minds. The mere fact that most adult teachers neither recognize nor feel any discomfort concerning the profound philosophical disorder of the curriculum they daily administer to children is itself a blatant indication of the unphilosophical character of the adult mind. Philosophy, one might say, is instinctively and naturally a concern of childhood, and only rather rarely and artificially still a matter of concern in adult life.

This in turn, of course, makes it difficult for adults to teach philosophy properly. The main mistake would be to teach philosophy as another academic subject, as a body of recognized problems, proposed solutions, and debates. The pupil would be expected to learn this up. This would, of course, miss the point entirely.[1] For what is needed is, in a sense, not the *teaching* of anything at all, but rather the encouraging of children themselves to engage in the activity of articulating and scrutinizing problems and their possible solutions. Furthermore, it would be vital to do this in an honest and open-ended way, there being no prohibitions on what problems can be discussed, what solutions considered. The nature of the universe, war, sex, death, power, money, politics, fame, pop stars, parents, school, work, marriage, the meaning of life, evolution, God, failure, drugs, love, suffering, happiness: whatever it is that the children find fascinating or disturbing, and want to discuss, deserves to be discussed. Where there are no known or no agreed answers, the teacher must acknowledge this. The teacher must readily acknowledge his or her own ignorance or uncertainties. The main task of the teacher will be to try to ensure that the children speak one at a time, that everyone gets to speak, and that those who are not speaking, listen. The teacher will also, of course, try to establish a spirit of generosity towards the ideas of others, while at the same time encouraging criticism and argument. The main object of the seminar is to enable children to discover for themselves the value of cooperative, imaginative, rational problem solving by taking part in it themselves. Only good, experienced teachers could hope to make a success of the philosophy seminar run along these lines.

The purpose of the seminar is not to promote mere *debate*. Argument is to be used as an aid to exploration and discovery: it is not to be used merely to trounce opponents or to win converts — as an excuse, that is, for intellectual duelling or bullying. The seminar must not be conducted in such a way that it amounts to overt or disguised *indoctrination* in some creed — however correct or noble the creed may be judged to be. Insofar as a creed is implicit in the seminar, it might be put like this: it is proper and desirable for people to resolve problems and conflicts in cooperative, imaginative, and rational ways. This creed

---

[1] This mistake is evident in current A-level philosophy syllabuses.

is itself open to discussion and critical assessment—along with all other political, religious, moral, economic, social, and philosophical doctrines. The problem of how to distinguish cooperative discussion from indoctrination deserves itself to be discussed when it arises. Again, the seminar is not *group therapy*. Its primary aim is not to solve the participants' urgent practical, personal problems (although it may occasionally and incidentally help to do this). Problems can be imagined and do not need to be lived. Ideas can be aired as possibilities, and do not need to be believed. Accounts of personal experience are welcomed when relevant to the discussion, but are not expected or demanded. The aim of the seminar is to explore possibilities, and not to reach decision about actions. Unanimity does not need to be sought.

It is nothing less than an educational scandal that seminars of this type are not a standard part of school and university life, available to everyone from the age of five years upwards. However, it is not just that there has been a general failure to organize all education around such a philosophy seminar. Worse still, there has been, and still is, a general failure even to see the vital need to do this. The very idea of the philosophy seminar for five-year-olds, as indicated here, has generally not been entertained. A major reason for this is that the proper purpose and character of philosophy, and of academic inquiry more generally, has long been, and still is, radically misunderstood, especially by academics themselves.

Academic inquiry is widely taken to have as its proper, basic intellectual task the improvement of expert, specialized knowledge and technological know-how. As long as academic inquiry is pursued and organized with this basic task in mind, the philosophy seminar, as depicted above, can scarcely form a normal, let alone a central, part of university work. Non-expert, non-specialized discussion of our problems of living—however imaginative, rational, cooperative, and potentially fruitful—cannot contribute to the acquisition of expert, specialized knowledge. Groups devoted to such discussion may amount to worthy debating societies, group therapy sessions, or Quaker prayer meetings: they cannot constitute standard *academic seminars*.

The fault here lies with the orthodox conception of academic inquiry. It is an intellectual and human disaster. When judged from the standpoint of improving specialized knowledge, orthodox academic inquiry must, it is true, be judged to be, on the whole, both rational and extraordinarily successful. But when judged from the more important and fundamental standpoint of improving human welfare, enhancing the quality of human life, academic inquiry must be judged to be grossly irrational and unsuccessful. In order substantially to improve the quality of human life on earth we need, amongst other things, to get rid of war, the threat of war, armaments whether nuclear, biological, chemical, or conventional, the extreme poverty of the third world, tyranny, exploitation and enslavement. Humanity needs to discover how to resolve its local and global conflicts and problems of living in more cooperatively rational ways. But cooperative *action* requires cooperative *discussion*. If academic inquiry is to devote itself, rationally and successfully, to promoting human welfare, then it must give priority to providing such cooperative discussion; it must, as a matter of absolute intellectual priority, (a) articulate our problems of living, and (b) propose and critically assess alternative possible solutions, possible cooperative *actions*. Problems of *knowledge* must be tackled in a subordinate way, scientific and technological research emerging out of and feeding back into the more fundamental concern with problems of *living*.

Contemporary academic inquiry, in giving priority to problems of knowledge over problems of living, fails to do what it most needs to do: create and promote a tradition of thinking devoted to resolving human conflicts and problems in cooperatively rational ways. In the absence of a general capacity to act cooperatively, the mere provision of knowledge and technological know-how can do as much harm as good, as the twentieth-century record of science and war, and the nuclear arms race, so horrifyingly exemplifies.

We urgently need, in brief, a new, more intellectually rigorous and humanly desirable kind of academic inquiry, one that gives priority to helping us realize what is of value in life, individually, locally, and globally. This new kind of inquiry gives intellectual priority to personal and social (or global) problems of living (rather than problems of knowledge) and endeavours to help us discover how to act, to live, in progressively more cooperatively rational ways, so that we achieve what is genuinely of value to us in the circumstances of our lives. The basic aim is to promote personal and social wisdom in life—wisdom being defined as the capacity to realize what is of value, for ourselves and others. Wisdom, so defined, includes, but goes beyond, knowledge and technological know-how. Given the existence of such a tradition of inquiry in the world, there is a real chance that humanity might learn how to make steady and substantial progress towards a generally happier state of affairs than that which we endure at present.

Once the academic community wakes up to the desperately urgent need to transform the academic enterprise in this way, so that its basic task becomes to promote not only *knowledge* but also personal and social *wisdom* in life, it will at once become blindingly obvious that the philosophy seminar, more or less as described above, does indeed need to be put at the heart of all inquiry and education, from primary school to university. Unfortunately, the academic community, despite being devoted to reason and innovation, is in many ways extremely conservative and highly resistant to change, especially when it comes to changing the overall aims and methods of inquiry. I am especially aware of this, having argued for some thirty years for the urgent need to change academic inquiry from knowledge to wisdom: so far I have seen few signs of change (see Maxwell, 1976a; 1980; 1984; 2000a; 2004a). If we wait for the scientists, scholars, and university administrators to wake up to what needs to be done, we may have to wait for ever. What we can do, and need to do, is begin with the five-year-olds. Professors may be past it, but five-year-olds are not.

The above was written long ago, in 1986, in complete ignorance of the philosophy for children movement. I then discovered Gareth Matthews' delightful little book *Philosophy for the Young Child* (1980), and as a result I laid aside this plea for philosophy for five-years-olds on the assumption that the matter was already satisfactorily in hand. Since then, philosophy for children has become a world-wide movement, and it might seem that this essay is redundant. This is not the case, for at least two reasons.

First, the philosophy for children movement seems to take for granted a thoroughly orthodox, analytic conception of philosophy, according to which philosophy is one discipline alongside others, concerned with puzzle solving and conceptual analysis. Given this conception of philosophy, it is difficult to see why philosophy should occupy a central and fundamental role in the curriculum. What is lacking is an awareness of the need to bring about a revolution in the aims and methods of academic inquiry as a whole, including philosophy and education, so that the basic aim becomes to acquire and promote wisdom, problems of living being put at the heart of the academic enterprise. Once one becomes aware of the need to bring about this revolution, it becomes clear that the philosophy seminar, along the lines I have indicated, ought to be central to all of education. The philosophy for children movement would, in my view, become more credible and cogent were it to join forces with the effort to transform inquiry as whole so that it takes up its proper task of promoting wisdom by rational means. Only within a genuinely rational kind of inquiry devoted to promoting wisdom can the philosophy seminar, as I have described it, come to have its proper place and role.

Second, in England the national curriculum all but prohibits the philosophy seminar as I have depicted it. Group discussion, listening and speaking, and problem solving are, it is true, all encouraged, and citizenship and personal, social, and health education are included. Furthermore, the curriculum for primary education may well be sufficiently flexible to permit something like the philosophy seminar to take place in individual schools. But there is, in the national curriculum, no hint that group discussion might feed into other parts of the curriculum, into science, history, or English. And when it comes to secondary education, the curriculum seems to be so rigidly constructed that it seems impossible that the philosophy seminar could get elbow room, let alone influence the rest of the curriculum.

We need to bring about a revolution in the national curriculum here in England, and we need a world-wide revolution in education and academia, so that the philosophy seminar comes to play a central role, for five- to ninety-five year-olds.

Chapter Two

What Philosophy Ought to Be

Introduction

The proper task of philosophy is to keep alive awareness of what our most fundamental, important, urgent problems are, what our best attempts are at solving them and, if possible, what needs to be done to improve these attempts. Unfortunately, academic philosophy fails disastrously even to conceive of the task in these terms. It makes no attempt to ensure that universities tackle global problems—global intellectually, and global in the sense of concerning the future of the Earth and humanity. Universities do not give sustained attention to global problems (due to specialization and giving priority to the pursuit of knowledge) and as a result violate three of the four most elementary rules of rational problem solving conceivable. Judged from the standpoint of helping humanity tackle global problems, universities as at present constituted betray reason and, as a result, betray humanity. Bereft of institutions of learning rationally designed to help us make progress towards as good and wise a world as possible, not surprisingly we fail to learn how to do it. This is the key crisis of our times. And it is, at root, a failure of *philosophy*. It is the failure of philosophy to keep alive rational exploration of global problems in universities, and in the public domain—a failure that can be traced back to the origins of modern philosophy in the 17th century. We urgently need a revolution in philosophy so that academic philosophers take up their proper task of promoting rational exploration of our fundamental, global problems.

1. What Philosophy Ought to Do

Philosophy is unique. There is no other academic discipline that has laboured for so long under such a massive misconception as to what its basic task ought to be.

The proper basic task of philosophy is to keep alive awareness of what our most fundamental, important, urgent problems are, what our best attempts are at solving them, and what the relative merits and demerits of these attempts are. A basic task is to articulate, and improve the articulation of, our fundamental problems, and make clear that there are answers to these problems implicit in much of what we do and think — implicit in science, politics, economic activity, art, the law, education, and so on — these answers often being inadequate and having adverse consequences for life and thought in various ways as a result.

Philosophy should also try to help *improve* our attempted solutions to our fundamental problems, by imaginatively proposing and critically assessing possible solutions, all the time making clear, where relevant, that different possible solutions have different implications for diverse aspects of life. As a result of improving our attempted solutions to our fundamental problems we may thereby contribute to the improvement of our lives, and help us make progress towards a good world.[1]

Even though these are the proper, fundamental tasks for philosophy, it hardly needs to be said that none of these tasks can be said to be the exclusive domain of philosophy or academic philosophers. Quite the contrary, a central task of

---

[1] This chapter not only argues that these should be the basic tasks of philosophy; it also, at the same time, seeks to make this kind of contribution to philosophy — that is, I try to practise what I preach. For earlier attempts of mine see Maxwell (1984; 1998; 2004a; 2007a,b; 2010a; 2014a). See also Chapter Five below.

philosophy is to stimulate as many people as possible to think about fundamental problems imaginatively and critically — that is, *rationally*. Philosophy is not to be characterized or delineated from other disciplines in terms of *who does it*, but rather in terms of *the fundamental character of the problems being tackled*, and perhaps *the value of the contribution in question*.[2]

What, then, are our fundamental problems? Our most fundamental problem of all, encompassing all others, can be put quite simply like this:

How can our human world, and the world of sentient life more generally, imbued with the experiential, consciousness, free will, meaning, and value, exist and best flourish embedded as it is in the physical universe?[3]

---

[2] There is an important, secondary, related task for philosophy: to get clear about the basic aims and methods of diverse, worthwhile, problematic endeavours — science, art, literature, politics, education, economic endeavour, the law, the media, living one's life, creating a good world, and so on — and to attempt to develop improved aims and methods, particular attention being given to providing frameworks for such improvement. The hope is, of course, that ideas for improved aims and methods will lead to actual improvements in the real world. This second task for philosophy is related to the first task, for ideas about how to improve attempted solutions to fundamental problems may also be ideas about how to improve aims of various endeavours, and vice versa. For an example of how this two-way interaction between ideas and aims can take place, see the discussion of *aim-oriented empiricism* in section 4 below. This important secondary task for philosophy is not discussed explicitly in this essay in what follows. It is however discussed in detail in Maxwell (1984; 1998; 2004a; 2007a; 2010a; and 2014a). For a summary, see Maxwell (2007b).

[3] I have devoted two books to articulating, and trying to help solve, this fundamental problem: see Maxwell (2001; 2010a). I there argue that this is our fundamental problem. See also Maxwell (1966; 1968a,b; 1984, ch. 10; 2000b; 2011a). For my discussion of that aspect of the problem concerned primarily with the *flourishing* of what is of value in our human world, see Maxwell (1984; 2004a; 2007a; 2014a); for summaries, see Maxwell (1980; 1992; 2000a; 2007b; 2009a). See also Chapter Five below.

Some will reject the idea that the ultimate reality behind the natural world is physical in character. For example, there are those who hold that the ultimate reality is God. In order not to exclude such views in an *a priori* fashion, as it were, we need a broader formulation of the above problem:

How can our human world… exist and best flourish embedded as it is in the real world?

I interpret the first formulation of this problem in such a way that it encompasses all of academic thought, from theoretical physics, mathematics, and cosmology, via the biological and technological sciences, to social inquiry and the humanities. It also encompasses all practical problems of living — problems facing individuals, groups, institutions, societies, nations, and humanity as a whole.[4]

The key idea of this conception of philosophy is that philosophy is concerned to help solve rationally our most *fundamental* problems. But what exactly does "fundamental" mean here?

We can perhaps say that problem P_1 is more fundamental than P_2 if solving P_1 also, at least in principle,[5] solves P_2, but not

---

[4] The moment it is accepted that philosophy has, as its basic task, to tackle fundamental problems, it is clear that philosophy education must be transformed. Instead of learning philosophy via the history of philosophy, rather one needs to plunge, from the outset, into the fundamental problem as it confronts us today, relevant background knowledge in physics, biology, climate science, social inquiry, and the humanities, politics, economics, and international affairs being acquired as one goes along. That the history of philosophy is the wrong way to learn philosophy becomes all the more obvious granted the points to be made below — namely, that much of philosophy in the past has been alienated from concern with our fundamental problems. For hints as to what is required see Chapter One, and Maxwell (2010a).

[5] Problems of quantum theory are more fundamental than problems of chemistry, and solving quantum theoretic problems may be regarded as providing solutions to problems of chemistry, but in general only in principle, not in practice, because in practice in order to solve chemical problems quantum mechanically, one needs to solve equations that

vice versa. This suffers from the disadvantage that "P_1 is more fundamental than P_2" in this sense might just mean that P_1 is more general. Can we distinguish "more fundamental" from "more general" — the former being stronger? It can be done like this. P_1 is more fundamental than P_2 if the solution to P_1 solves P_2, but not *vice versa*, and the solution to P_1 is unified or coherent in some significant, substantial sense of these terms, and not just a jumble of disconnected items. An example of a unified or coherent solution is a unified physical theory that solves a range of problems in physics.[6]

Granted this conception of the basic task of philosophy, it at once becomes clear that philosophy in the university has, as an elementary obligation, to ensure that sustained thinking about our fundamental problems and how to solve them goes on in an influential way within academic inquiry. This is, indeed, a basic requirement for academic inquiry to be rational. Four elementary, almost banal, rules of reason are:

(1) Articulate, and seek to improve the articulation of, the basic problem to be solved.

(2) Propose and critically assess possible solutions.

(3) If the basic problem to be solved proves intractable, specialize. Break the basic problem up into subordinate problems. Tackle analogous, easier-to-solve problems, in an attempt to work gradually to the solution to the basic problem to be solved.

(4) But if one engages in specialized problem solving in this way, make sure that specialized and basic problem solving interact, so that each influences the other (since otherwise specialized problem solving is likely to become unrelated to the basic problems we seek to solve).

---

often cannot be solved (associated with interactions of many complex molecules, for example).

[6] For what it means to say of a physical theory that it is unified see Maxwell (1998, chs. 3 and 4; 2004a, appendix, section 2; 2007a, ch. 14, section 2; 2013a, section 4).

Sustained thinking about what we may call "global" problems — global intellectually, and global in the sense of encompassing the earth and humanity as a whole — must go on in universities in a way that influences, and is influenced by, more specialized research if rules (1), (2), and (4) are to be put into practice, and academic inquiry is to meet elementary requirements for rationality. Philosophy as sustained thinking about our fundamental problems and how to solve them must be an integral, influential part of academia if academia as a whole is to be rational. A quite basic task for philosophy, then, is to ensure, as a bare minimum, that universities are organized in such a way that each university has a big, prestigious Seminar or Symposium, open to all at the university from undergraduate to vice-chancellor, which meets regularly to explore global problems in a sustained way, and in a way that is capable of influencing, and being influenced by, more specialized research.

From what I have said so far, one would expect such global seminars to be commonplace in universities around the world.

I know of no university anywhere that has such a global seminar.[7]

Academic philosophy has failed dismally to create such a global seminar in the university. Even worse, it has made no attempt to do so. Worse still, academic philosophy has failed almost entirely to take on the task I have indicated above — the task of keeping alive awareness of what our most fundamental problems are (as a bare minimum).

---

[7] Attempts have been made, recently, however, in a few universities to introduce sustained interdisciplinary research into global problems: see, for example, the Grand Challenges Programme at my own university, UCL, at www.ucl.ac.uk/research/grand-challenges. On the UCL website, www.ucl.ac.uk/, under "Research", there appears "The Wisdom Agenda" which, if clicked on, reveals a document of the same title which may be downloaded. There is here an input from my own work. (Websites accessed 22 July 2013.) For an indication of recent changes in academia in the direction I argue for in this essay see Maxwell (2009b).

Academic philosophy today does not even recognize, as a fundamental problem of the discipline: *What kind of inquiry can best help us realize what is of value in life?* or, to quote the title of an article of mine, *What kind of inquiry can best help us create a good world?* (Maxwell, 1992).

2. A Fundamental Failure of Philosophy

Academic philosophy does discuss some technical, conceptual puzzles associated with the fundamental problem I have indicated above. There is discussion of puzzles associated with the mind/body problem, free will and determinism, the question of whether physical theory can be interpreted "realistically" as postulating unobservable physical entities such as electrons and quarks, and discussion of some related conceptual issues having to do with such things as knowledge, perception, reason, action, the good, justice, what is of value. But the basic tasks for philosophy that I have indicated above are just not done.[8]

The consequences of this abysmal failure of academic philosophy to do what it most needs to do are dire indeed. The outcome is that academia as a whole fails both reason and humanity. The failure of academic inquiry to give an important role to the sustained exploration of global problems within the university means that academia violates three of the four most elementary rules of reason that one can think of—rules (1), (2), and (4). Rule (3) is of course put splendidly into effect in all our universities. Disciplines splintering again and again and again into ever more specialized subordinate disciplines is one of the most striking features of the university today.[9] But the failure to

---

[8] My claim is that academic philosophy fails to put our fundamental problem, as I have articulated it above, at the centre of the discipline. I do not want to suggest that no philosopher has ever discussed the problem. It is, for example, a theme of Whitehead (1932).

[9] There may well be good intellectual reasons for specializations, as rule (3) indicates. But it may come about for entirely non-intellectual, spurious, and reprehensible reasons, having to do with promoting

tackle fundamental problems in a sustained and influential way means that rules (1) and (2) are violated, which in turn means that rule (4) is violated as well.[10]

This wholesale, structural breakdown of rationality is no mere formal matter. It has dire consequences for humanity. This long-standing structural irrationality of academia is in part responsible for the genesis of our current global problems, and our incapacity to resolve them effectively and wisely. People die as a result.

Consider some of the most serious global problems that face humanity today: rapid growth in the world's population, the lethal character of modern war and terrorism, immense differences in wealth and power around the globe, destruction of natural habitats and rapid extinction of species, pollution of earth, sea, and air—and, grimmest of all, perhaps, the impending disasters of climate change.

What would resolve these problems in such a way that the outcome is a more peaceful, just, equable, democratic, sustainable world—a world in which we all have good chances of leading lives of value? Certainly relevant scientific knowledge, understanding, and technological know-how are essential. But these problems would be resolved fundamentally not by knowledge or technological know-how but by appropriate *actions*. It is what we do, or refrain from doing, not what we know that enables us to realize what is of value in life (except when knowledge is of value in itself). Even when scientific knowledge and technological know-how are relevant, as they are in medicine or agriculture for example, it is always what this knowledge enables us to *do* that leads to the achievement of what is of value, not the knowledge in itself.

---

careers, research groups, and special interests. Creating a new speciality, with its own journals and jargon, can do much for academic careers.

[10] That academia should include sustained rational discussion of fundamental problems was argued long ago in Maxwell (1980). See also Maxwell (2010a).

Thus, in order to solve our global problems we need to discover how to *do* what needs to be done to resolve them. We need, fundamentally, to discover how so to act, to live, that we tackle our global problems in increasingly effective, intelligent, and humane ways.

We need to *learn* how to do it. We need to *learn* how to develop and implement new political programmes, new policies, new economic strategies, new ways of living. We need to improve our institutions, our trading relations, our laws and customs, our politics, our media, the content of our communications. Above all, I would suggest, we need to learn how to tackle our global problems in increasingly cooperatively rational ways.

We are confronted, then, by an immense task of learning, and that, in turn, means that it is vitally important that our *institutions of learning* — our universities and schools — are properly organized, structured, and devoted to helping us learn what we need to learn. Our universities need to be organized and devoted, fundamentally, to helping us learn how so to act, to live, that we progressively resolve our conflicts and problems of living, including our global problems, in such ways that, with increasing success, we come to realize what is genuinely of value in life.

In short, granted that the basic aim of academia is to help promote human welfare, help people realize what is of value in life, the problems that need to be tackled are, fundamentally, problems of living, problems of action in the real world and not, primarily, problems of knowledge. A basic academic task must be to promote cooperatively rational tackling of problems of living in the great social world beyond the confines of the university. Universities cannot of course decide for the rest of us what our problems of living are and what we need to do about them. Their job is to propose, to argue, to critically assess, to promote awareness of what our problems may be, and what may be our options. And to learn from, and spread awareness of, good solutions in practice wherever they are to be found in the community. One might think of universities as a kind of people's civil service doing openly for the public what actual civil services are supposed to do, in secret, for governments. Universities need just sufficient power to retain their independence from government, industry, the media, the military, public opinion, but no more.

A kind of academic inquiry well-designed to help promote human welfare, in short, must, as a matter of absolute intellectual priority (1) articulate problems of living (including global problems), and (2) propose and critically assess possible solutions—possible actions, policies, political programmes, economic strategies, ways of life. It must also, of course, (3) engage in specialized scientific and technological problem solving, but must, at the same time, (4) ensure that fundamental and specialized problem solving influence each other, so that fundamental problem solving is informed of the results of specialized research, and specialized research retains its relevance to our fundamental problems of living.

If universities were designed in this way around the world, there might be some hope that we would gradually learn how to resolve our grave global conflicts and problems in increasingly cooperatively rational ways, thus gradually making progress towards a better, wiser world. But universities are not remotely designed or organized in this vitally necessary way. From the past we have inherited the idea that academia must devote itself, in the first instance at least, to the pursuit of knowledge. First, knowledge is to be acquired; then, in a secondary way, it can be applied to help solve social problems. The vitally necessary task of tackling problems of living imaginatively and critically is excluded from the intellectual domain of inquiry, or pushed to the periphery and marginalized. What universities most need to do to help humanity learn how to make progress towards as good a world as possible is not done at all, or is only done in a severely restricted fashion, and certainly not as the central, primary concern.

This is a failure of philosophy. It is the failure of philosophy to establish that universities need to give sustained attention to fundamental problems in order to meet elementary requirements of rationality, and in order to serve the best interests of humanity. It is the failure of philosophy even to conceive of the need to do this.

The outcome of this failure is that, instead of helping to solve global problems, universities have, if anything, actually helped to create and intensify these problems.

It is all too rarely appreciated that modern scientific knowledge and technological know-how have made all our current global problems possible. Much of great benefit has of course come from science and technology. They have made the modern world possible. But in making possible modern industry and agriculture, modern medicine and hygiene, modern transport and armaments, they also made possible all the global problems indicated above: the explosive growth in the world's population, vast inequalities in wealth and power around the world, the lethal character of modern war, climate change, and the rest.

There is a sense, indeed, in which science and technology may be said to be the *cause* of these things. It will be said at once that it is not *science* that is the cause of these global problems but rather the things that we *do*, made possible by science and technology. This is obviously correct. But it is also correct to say that scientific and technological progress *is* the cause. The meaning of "cause" is ambiguous. By "the cause" of event E we may mean something like "the most obvious observable events preceding E that figure in the common sense explanation for the occurrence of E". In this sense, human actions (made possible by science) are the cause of such things as people being killed in war, destruction of tropical rain forests. On the other hand, by the "cause" of E we may mean "that prior change in the environment of E which led to the occurrence of E, and without which E would not have occurred". If we put the 20th century into the context of human history, then it is entirely correct to say that, in this sense, scientific-and-technological progress is the cause of our distinctive current global disasters: what has changed, what is new, is scientific knowledge, not human nature. Give a group of chimpanzees rifles and teach them how to use them and in one sense, of course, the cause of the subsequent demise of the group would be the actions of the chimpanzees. But in another obvious sense, the cause would be the sudden availability and use of rifles—the new, lethal technology. Yet again, from the standpoint of theoretical physics, "the cause" of E might be interpreted to mean something like "the physical state of affairs prior to E, throughout a sufficiently large spatial region surrounding the place where E occurs". In this third sense, the sun continuing to shine is as much a part of the cause of war and pollution as human action or human science and technology.

In short, if by the cause of an event we mean that prior change which led to that event occurring, then it is the advent of modern science and technology that has caused all our current global crises. It is not that people became greedier or more wicked in the 19th and 20th centuries; nor is it that the new economic system of capitalism is responsible, as some historians and economists would have us believe. The crucial factor is the creation and immense success of modern science and technology.

Many blame science for our problems. But that misses the point. It is not science that is at fault, but rather science *dissociated from a more fundamental concern with our problems of living and what to do about them*. The fault lies with our failure to develop a kind of inquiry, sketched above, rationally designed and devoted to helping us learn how to solve our problems of living, realize what is of value to us in life. The fault lies not with science but with philosophy.[11]

---

[11] Philosophers do not even have the excuse that the argument for the urgent need to transform academic inquiry has not been spelled out in the literature. On the contrary, I have spelled out the argument in great detail again and again for decades: see Maxwell (1976a; 1984; 2004a; 2010a; 2014a). For summaries of the argument see Maxwell (1980; 1992; 2000a; 2003; 2007b; 2008a,b; 2009a; 2012b; 2013a).

3. How Philosophy Came to Fail so Drastically

How and why did philosophy come to fail so drastically? Once upon a time, it is clear, philosophy had no inhibitions at all about tackling fundamental problems. What kind of universe is this? How did we come to be? What is of most value in life? What kind of social world should we strive to create? The ancient Greek philosophers tackled these fundamental problems in stark, bold terms: this is the case, for example, of Thales, Anaximander, Heraclitus, Democritus, Socrates, Plato, Aristotle. Early modern philosophers did this too: Descartes, Leibniz, Locke, Hobbes, Spinoza, Kant. We need to remember, indeed, that modern science began as an extraordinarily successful outgrowth of philosophy. The creators of modern science, Kepler, Galileo, Hooke, Boyle, Huygens, Newton, and their contemporaries all thought of themselves as engaged in philosophy — in natural or experimental philosophy. And the basic task of natural philosophy was to improve our answers to the fundamental philosophical problem: What kind of universe is this? Kepler, Galileo, Descartes, Boyle, Huygens, Leibniz, and other natural philosophers of the time did not hesitate to put forward their conjectures about the nature of the universe along with proposed laws and theories about more specific phenomena such as terrestrial and astronomical motion, sunspots, the tides, light, gases, and so on.

What happened? What caused philosophy to abandon tackling fundamental problems?

It all goes back to the triumph of Newtonian physics and, in particular, associated with this, the triumph of Newton's conception of scientific method.

In his *Principia*, Newton claimed to derive his law of gravitation from the phenomena by induction without framing hypotheses. He claims to do this as follows. First, from his three basic laws of motion, Newton proves mathematically a great number of theorems which concern, amongst other matters, bodies that move along elliptical and parabolic paths, and bodies that move under the influence of a force directed towards a fixed centre. He proves, for example, that a body in motion under the influence of a force directed towards a fixed centre that varies inversely as the square of the distance will travel along an elliptical path. Newton then formulates four "rules of reasoning in philosophy". These specify, in somewhat different ways, how universal laws may be arrived at by induction from observed regularities, without resort to metaphysical or philosophical hypotheses. Newton then formulates six phenomena, six astronomical regularities. These concern the manner in which the moons of Jupiter, Saturn, and Earth in their motions around their respective planets, and the motions of the six inner planets in their motions around the sun, observe Kepler's laws of planetary motion. From these phenomena, Newton then goes on to derive by induction his universal law of gravitation, invoking during the course of this inductive derivation his mathematical theorems, and his four rules of reasoning.[12]

For some years after the publication of Newton's *Principia* in 1686, natural philosophers fell into two camps. On the one hand those in England supported Newton, while those on the Continent, by and large, supported Descartes. As Voltaire put it decades later in his *Lettres Philosophiques*:

> A Frenchman arriving in London finds things very different, in natural science as in everything else. He has left the world full, he finds it empty. In Paris they see the universe as composed of vortices of subtle matter, in London they see nothing of the kind... For your Cartesians everything is moved by an impulsion you don't really understand, for Mr. Newton it is by gravitation, the cause of which is hardly better known.[13]

The astonishing predictive and explanatory success of Newtonian theory, together no doubt with his claim to have derived his universal law of gravitation from the phenomena by induction without appealing to metaphysical hypotheses, led eventually

---

[12] Newton (1962).
[13] Voltaire (1980, p. 68).

to the downfall of Cartesian physics and cosmology, and the triumph of Newton. And along with the victory of Newtonian physics came the victory of Newtonian methodology. Descartes' somewhat rationalistic, *a priori* methods of "clear and distinct ideas" fell into disfavour. Instead, after the immense success of Newtonian physics, natural philosophers had, it seemed, for the first time in history, a clear way forward. What one had to do in order to acquire reliable knowledge of nature was to put Newton's rules of reasoning into practice. First, discover regularities in the natural world by means of observation and experiment. Then, apply Newton's rules of induction to arrive at universal laws and theories. Philosophical and metaphysical speculation no longer had any role whatsoever in natural philosophy—or in "natural science" as it came subsequently to be called. Scientists could ignore philosophy, and exploit Newton's extraordinarily successful empirical methods. Thus gradually after Newton, natural philosophy was reborn as science.[14]

A gulf opened up between science and philosophy. Scientists came to feel that they could safely ignore philosophy, as irrelevant to the task of improving scientific knowledge of the natural world by means of the established methods of natural science bequeathed to them by Newton. And philosophy for its part participated in the creation of this gulf by failing to produce anything of interest or of use to the new science. This failure stemmed from a more basic failure to solve fundamental problems thrown up by the new natural philosophy, and the new science. As a result, philosophy became more and more remote from science. The natural philosophy of Galileo, Kepler, Descartes, and even Newton, broke up into natural science on the one hand, philosophy on the other.[15] So vast and decisive is this gulf that, in a wholly anachronistic way, it is today pro-

---

[14] The term "scientist", however, only came into use in the middle of the 19th century after it was introduced by William Whewell in 1834.
[15] Elsewhere I have argued that we need to recreate natural philosophy: see Maxwell (1984 or 2007a, ch. 9; 2012a).

jected back into the past, so that nowadays we divide up 16th- and 17th-century natural philosophers, quite artificially, into two camps: the scientists (Kepler, Galileo, Huygens, Newton), and the philosophers (Descartes, Locke, Leibniz, Hobbes, Spinoza). They would not have seen themselves in this fashion. They would have held themselves to have been natural philosophers without exception.

Philosophy failed to solve two absolutely fundamental problems created by the new natural science, namely:

1. How is it possible for science to establish universal laws and theories by means of inductive inference from evidence?

2. If the universe really is more or less as modern science seems to tell us it is, how can our human world exist, imbued as it seems to be with colours, sounds, smells, tactile qualities as we experience them, and with consciousness, free will, meaning, and value? If the universe is, in the end, more or less as depicted by physics, does not that mean that the world as we experience it is almost entirely an illusion?

It is the failure of modern philosophy to solve these two fundamental problems that accounts for its progressive alienation from its basic task: to keep alive awareness of our fundamental problems. I take these two problems in turn in the next two sections.

4. The Problem of Induction

Problem 1 arises because, however much evidence there is in support of a physical theory, Newtonian theory let us say, or quantum theory — however severely tested the theory may be — endlessly many rival theories can be concocted which fit all the available evidence just as well as the given theory. We can concoct endlessly many such rivals by modifying the given theory in wholly ad hoc ways so that each new theory differs from the initial theory only for some as yet unobserved phenomenon — for example, some phenomenon that lies in the future.[16] The

---

[16] See Maxwell (1998, ch. 2, section 7; 2013a, section 3).

problem was formulated in a particularly striking way by David Hume.[17] It led Immanuel Kant to ask "How is science possible?"[18] Ever since, philosophers have struggled to answer Kant's question, and have failed.[19] Nothing could highlight more dramatically the difference between science and philosophy. Whereas science goes from strength to strength, philosophy goes backwards. It is reduced to trying to work out how any theoretical knowledge in science can be achieved at all. Far from contributing to the success of science, for philosophy it is this very success that poses the problem. Philosophy has, it seems, nothing fruitful or helpful to contribute to science at all. And this tends to be the opinion of scientists themselves. Some years ago John Ziman, a physicist, wrote, "the Philosophy of Science... [is] arid and repulsive. To read the latest symposium volume on this topic is to be reminded of the Talmud, or of the theological disputes of Byzantium".[20] More recently Steven Weinberg declared: "From time to time... I have tried to read current work on the philosophy of science. Some of it I found to be written in a jargon so impenetrable that I can only think that it is aimed at impressing those who confound obscurity with profundity... [O]nly rarely did it seem to me to have anything to do with the work of science as I knew it... I am not alone in this; I know of *no one* who has participated actively in the advance of physics in the post-war period whose research has been significantly helped by the work of philosophers."[21] Recently, Stephen Hawking pronounced that "philosophy is dead".[22] Given the apparent impotence of philosophy to be of any help to science, these comments are hardly surprising.

---

[17] See Hume (1959).
[18] Kant (1953, pp. 52–89).
[19] For references to failed attempts at solving the problem see Kyburg (1970), Swain (1970), Howson (2000).
[20] Ziman (1968, p. 31).
[21] Weinberg (1993, pp. 133–4).
[22] Hawking and Mladinow (2010, ch. 1).

As it happens, the problem of induction has been solved, and a philosophy of science has been put forward that would, if put into scientific practice, be genuinely fruitful for science.[23] By and large, this solution has been overlooked by both philosophers and scientists.[24]

In order to solve the problem of induction, we need first to follow Karl Popper, and acknowledge scientific theories cannot be verified empirically; they can only be falsified.[25] We then need to appreciate that theories in physics have to satisfy *two* requirements to be accepted. They must be sufficiently empirically successful; and they must be sufficiently *unified* (that is, they must postulate near enough *the same* laws for the range of phenomena to which the theory applies).[26] We then need to appreciate that persistent acceptance of (more or less) *unified* theories even though endlessly many empirically more successful disunified rivals can easily be concocted means that physics makes a big metaphysical assumption about the universe: there is some kind of underlying dynamic unity in nature.[27] Then it needs to be appreciated that this assumption, because of its substantial, influential, and highly problematic character, needs to be represented in the form of a hierarchy of assumptions (and associated methods), assumptions becoming less and less substantial, and more nearly such that they must be true for science, or the pursuit of knowledge, to be possible at all. At each level in the hierarchy, that assumption is adopted which best accords with the assumption above, and leads to the most empirically progressive research programme, or offers the best promise of leading to such a programme. Assumptions are subjected to

---

[23] See Maxwell (1974; 1998; 2004a, especially appendix; 2005; 2006; 2011b; 2013a; and especially 2007a, ch. 14).

[24] One scientist and philosopher of science who has not overlooked it is Alan Sokal: he supports aim-oriented empiricism (personal communication). See also Longuet-Higgins (1984).

[25] Popper (1959; 1963, chs. 1, 10, and 11).

[26] See note 6 on page 15.

[27] See note 23.

sustained criticism, alternatives being developed and assessed, in an attempt to improve the assumptions that are adopted, criticism being concentrated where it is likely to be most fruitful, near the bottom of the hierarchy. This *aim-oriented empiricist* conception of physics (as I have called it) enables us to improve assumptions and methods—aims and methods—as scientific knowledge and understanding improve. There is something like positive feedback between improving scientific knowledge, and improving assumptions and methods, improving knowledge about how to improve knowledge—the nub of scientific rationality, according to this view.[28]

Not only does aim-oriented empiricism solve the problem of induction. Putting it explicitly into practice would have fruitful implications for science.[29] The centuries-long scientific poverty of philosophy comes to an end. In making explicit implicit metaphysical assumptions of physics, and in providing a framework of relatively unproblematic, fixed assumptions (high up in the hierarchy), aim-oriented empiricism provides a framework for the *improvement* of more substantial and problematic metaphysical assumptions, lower down in the hierarchy.[30] Aim-oriented empiricism provides physics with a rational, if fallible

---

[28] See note 23.
[29] Science has made progress because it has put aim-oriented empiricism into practice. But this has been done in only an implicit, unacknowledged, furtive, and partial fashion because scientists have sought to make science conform to their conviction that *standard empiricism* ought to be implemented—a bad philosophy of science that holds that empirical and simplicity considerations alone decide what theories are accepted and rejected in science, *no substantial claim about the universe being accepted as scientific knowledge independent of evidence*. The attempt to make science conform to standard empiricism has checked the explicit implementation of aim-oriented empiricism in practice, and as a result has subverted scientific progress somewhat. Rejection of standard empiricism, and the explicit and thoroughgoing acceptance and implementation of aim-oriented empiricism in its stead, would be of great benefit to science: see Maxwell (1998, ch. 1; 2004a, ch.2 and appendix; 2008a).
[30] A point stressed in works referred to in note 23 on the previous page. See especially Maxwell (2004a, appendix, section 5).

and non-mechanical, method for the discovery of revolutionary new theories.[31] And aim-oriented empiricism clarifies what it means to say of a physical theory that it is *unified*, and provides a partial ordering of theories with respect to degrees of unity.[32] Furthermore, aim-oriented empiricism has implications throughout natural science, and not just for theoretical physics.[33]

Aim-oriented empiricism transforms science, philosophy, and the relationship between the two.[34] Philosophy of science, insofar as it is about what are, and ought to be, the aims and methods of science, becomes an integral part of science itself, within the framework of aim-oriented empiricism.[35] And science, in a sense, ceases to be science and becomes much more like natural philosophy as it was in the time of Newton. Metaphysics, methodology, epistemology, philosophy all become a vital, integral part of science itself, as in Newton's time. The great divide between science and philosophy, inherited from Newton, is no more—or would be no more, if aim-oriented empiricism were to be adopted.[36]

But as long as the untenable, orthodox view is taken for granted that evidence alone determines what is accepted in science, philosophy will continue to be largely irrelevant to science. The chances are that philosophers of science will continue to ask despairingly the Kantian question "How is scientific knowledge possible?", and will not contribute to attempts to

---

[31] See Maxwell (1993a, pp. 275–305; 1998, pp. 219–23; 2004a, pp. 34–9 and 191–205).
[32] See note 6 on page 15.
[33] See Maxwell (2004a, pp. 39–67; 2008a).
[34] Elsewhere, I have argued that aim-oriented empiricism needs to be generalized to form *aim-oriented rationality*, a conception of rationality designed to help us improve problematic aims as we act. Applied to academic inquiry, it leads to the conclusion that knowledge-inquiry needs to be transformed so that it becomes *wisdom-inquiry*—a kind of inquiry designed to help humanity realize what is genuinely of value in life, make progress towards as good and wise a world as possible: see works referred to in note 11 on page 22.
[35] See Maxwell (2004a, pp. 39–47).
[36] See Maxwell (1998; 2004a, pp. 47–51; 2008a; 2012a; 2013a).

solve the fundamental problems tackled by science, and created by our scientific knowledge and understanding.

One day, perhaps, scientists may come to look favourably on aim-oriented empiricism. Even philosophers may eventually take note of the view. Then natural philosophy might be recreated, and academic philosophy might again begin to take up its proper tasks.

5. The Human World/Physical Universe Problem

Associated with the birth of what we now call modern science (but was then called natural philosophy), there was a revolution in philosophy. Aristotelianism was rejected, and atomism was adopted instead. But atomism creates a profound problem concerning the existence and value of the human world. If the universe really is made up solely of atoms that interact in accordance with precise laws, and are bereft of all experiential qualities such as colours, sounds, and smells, how can the world exist as we experience it, full of colours, sounds, and smells? How can our inner experiences exist, our thoughts and feelings, our states of consciousness? How can we be responsible for our actions — how can we have free will? How can human life have any meaning or value?

Atomism as adopted by Galileo (1564–1642), Descartes (1596–1650), or Huygens (1629–1695) is very different from the view of the universe adopted by physicists today. But the dramatic changes in our conception of the physical universe that have come about since the 17th century have not in themselves had much impact on the problem just indicated — *the human world/physical universe problem* (HW/PhU problem) as it may be called. What is common to our view of the universe today and the atomism of the 17th century, a doctrine that may be called *physicalism*, can be put like this: the universe is made up solely of one kind of physical entity (perhaps one entity), that interacts in accordance with precise (perhaps probabilistic) physical law. (Aim-oriented empiricism tells us that the basic physical entity, some kind of physical field pervading all of space and time, interacts with itself in accordance with a unified pattern of physical law.) It is physicalism that poses the human world/physical universe problem.

This problem posed by science, posed by the metaphysical view of the universe associated with modern science, is a *philosophical* problem—indeed *the* philosophical problem *par excellence*, as I proposed at the beginning of this essay. And it has, in a way, been central to philosophy since Galileo and Descartes. But attempts at solving the problem over the centuries have been disastrous failures. And it is this long-standing failure that has led much of philosophy to become remote from science, to become alienated from its basic problems and tasks, and to become lost in esoteric trivialities. The degeneration of philosophy has been the outcome.

An early and famous attempt at the solution is due to Descartes.[37] Cartesian dualism divides reality into two realms: the physical universe; and the world of minds. Physicalism is correct about the material world. Everything that physics leaves out, the sensory qualities we experience, are to be scooped up from the world around us and tucked into our minds. Minds are associated with, distinct from, but in interaction with, living brains of persons.

Cartesian dualism is a brilliant attempt at the solution to the HW/PhU problem. But it faces lethal problems. There is the problem of the wild implausibility of these mysterious entities, conscious minds, somehow being associated with physical processes going on in our brains, but utterly distinct from them. There is the problem of the *interaction* between brain and mind. Mind must interact with brain if we are to have free will, but such an interaction would mean that physical processes occur in our brains which cannot even in principle be explained physically. Cartesian dualism must postulate persistent, minute, poltergeist events in the brain. Physicalism is violated. But by far the most serious problem confronting Cartesian dualism is that it implies (or seems to imply) that it is impossible for us to

---

[37] Descartes (1949).

acquire any knowledge of the physical world around us. The world we experience, what we see, hear, touch, taste, smell, does not exist. It is all in the mind. How then can we experience any aspect of the physical world? We are locked inside our minds. And physics, applied to the processes of perception, seems to confirm this. Light enters our eyes, which causes physical processes to travel up our optic nerve to our brain, and then we have the experience of seeing, a mental event remote from, and utterly different from, its external cause in the physical world.

Given that Cartesian dualism faces these horrendous problems, the sensible, rational thing to do would be to reject the doctrine, return to the original problem that it sought to solve, the HW/PhU problem, and think again. If that had occurred, academic philosophy might not be in the dire state that it is in today. But that is not what occurred. Instead, something far more paradoxical took place. Many, perhaps most, subsequent philosophers did reject Cartesian dualism. But they accepted many implications of Cartesian dualism. They struggled to solve problems bequeathed to them by Cartesian dualism. And as a result, philosophy became more and more removed from, and irrelevant to the problems posed by, science. What philosophy after Descartes singularly failed to do was to return to the fundamental problem Descartes tried, and failed, to solve: the HW/PhU problem. Even worse, philosophical doctrines came to prevail which, once accepted, made it impossible even to articulate the HW/PhU problem.

Descartes led to Locke (1632-1704). Locke, ostensibly much more of an empiricist than Descartes, held that all our ideas stem from sense impressions, and was more doubtful than Descartes about the nature of physical entities and our capacity to acquire knowledge about them. But essentially, Locke accepted Descartes' dualism.[38] Locke led to Berkeley (1685-1753). Berkeley pointed out that if all we ever experience is our

---

[38] Locke (1961).

inner sensations, then we can never perceive external objects, and we can have no reason whatsoever to suppose that they exist. We can have no knowledge whatsoever of the material world. The physical universe disappears. All that science is about vanishes. To be is to be perceived. There is only the world of mind, of immediate sensation and experience.[39] Berkeley led to Hume (1711–1776). Hume pointed out that, if all our ideas stem from our sense impressions, then any idea which cannot be traced back to sense impressions cannot be meaningful. It cannot be an idea at all. But ideas about things external to us, the material universe, and what causes one state of affairs necessarily to result in another all belong to this category of meaninglessness. We cannot even have meaningful ideas about a physical universe of which we can have no direct experience. Science interpreted to be about the material world is not just impossible. It is meaningless.[40]

Hume led to Kant (1724–1804). And Kant endorsed and intensified Hume's ferociously sceptical attitudes towards the material world, the entire domain of natural science. Kant thought that the material world — or the *noumenal* world as he called it — does exist, but he held firmly that nothing whatsoever can be said about it, except that it exists. The subject of science, insofar as it is the material world, has been removed entirely from human reach.[41]

Kant is a paradoxical character. He took science very seriously, and even contributed to it.[42] Nevertheless, according to Kant, science is about the phenomenal world, the world of experience, not the real world, the material world, which is, for Kant, unknowable.

---

[39] Berkeley (1957).
[40] Hume (1959).
[41] Kant (1950).
[42] Kant contributed to the nebula hypothesis concerning the origins of the solar system: according to this hypothesis, a mass of particles surrounding the sun gradually coalesced to form the planets.

The outcome of this progression in philosophy, from Locke to Kant, is that the fundamental problem of philosophy, the HW/PhU problem, cannot be formulated. That which sets the problem, the physical universe, has been intellectually annihilated, or at least cast into the realm of the utterly unknowable. Instead of Cartesian dualism and its implications being firmly rejected, the implication concerning the impossibility of knowing anything about the physical world by means of experience is firmly adopted, and as a result it becomes impossible even to formulate the HW/PhU problem.

Kant's philosophy, famous for its obscurity, led to a great upsurge of obscure work in metaphysics, often idealist, anti-rationalist, and indifferent to, if not hostile towards, natural science. Kant led on, unwittingly, to Fichte (1762-1814), Schelling (1775-1854), Schleiermacher (1768-1834), Hegel (1770-1831), Schopenhauer (1788-1860), Husserl (1859-1938), and Heidegger (1889-1976). Bombastic metaphysics became all the rage, spreading even to Britain with the work of T.H. Green (1836-1882), F.H. Bradley (1846-1924), and J. McTaggart (1866-1925), and to France with existentialism and the work of Sartre (1905-1980) and Merleau-Ponty (1908-1961). The anti-scientific and idealist character of this body of post-Kantian work again made it impossible even to formulate the basic problem of philosophy, the HW/PhU problem.

Inevitably, a reaction set in. G.E. Moore (1873-1958) did much to initiate it by criticizing some of the outlandish assertions of the metaphysicians in the name of common sense.[43] Bertrand Russell (1872-1970) along with his one-time student Ludwig von Wittgenstein (1889-1951) contributed to the reaction by emphasizing that the world is made up of *facts*. They propounded a doctrine called *Logical Atomism*. There is an element of irony in this being a part of a movement against metaphysics in that the doctrine has itself a distinctly metaphysical air about it, especially in the hands of Wittgenstein.

---

[43] Moore (1959).

Logical Atomism holds that the world is made up of *atomic facts* —facts that are logically independent of one another.[44] One problem this doctrine faced was that no one could come up with a single convincing example of an atomic fact. There are good grounds for holding that there are none—as a glance at Maxwell (1968a) might convince one. Facts in the real world tend to be logically related to one another. This is especially true of facts about the physical universe.

Russell also contributed to the anti-metaphysical movement by helping to establish the view that the proper job of philosophy is *analysis*—logical, philosophical, or conceptual. And Russell produced what was later taken to be a paradigmatic case of philosophical analysis. This holds that "The King of France is bald" is to be analysed to assert "There is a man who is at present King of France; there is only one such man; and he is bald".[45]

Vienna in the 1930s then spawned a movement dedicated to the celebration of science and the annihilation of metaphysics once and for all. This movement is called *Logical Positivism*, and its members included Moritz Schlick (1882–1936), Rudolf Carnap (1891–1970), Carl Hempel (1905–1997), Otto Neurath (1882–1945), Hans Reichenbach (1891–1953), Friedrich Waismann (1896–1959), Herbert Feigl (1902–1988), and Philipp Frank (1884–1966). Wittgenstein was a sort of aloof figurehead. According to Logical Positivism, the meaning of a proposition is given by the method of its verification. All meaningful propositions fall into two classes, empirical and analytic. Empirical propositions are verified by an appeal to evidence, analytic ones by an appeal to the meaning of constituent terms, as when we convince ourselves that "All bachelors are unmarried" is true in virtue of the meaning of "bachelor" and "unmarried". Analytic propositions can be established with certainty but assert nothing

---

[44] Russell (1956); Wittgenstein (1960).
[45] Russell (1905).

about the world. Only propositions verified empirically make assertions about the world.

Metaphysical propositions, however, are put forward as being about the world that have been proved by reason alone. But this is not possible. Such propositions are neither empirical nor analytic. Hence they are all meaningless.[46]

Logical Positivism faced the dreadful problem, however, that scientific laws and theories cannot be conclusively verified either, and thus are all meaningless too. The Logical Positivists struggled to formulate a version of the verification principle that included as meaningful only that which they wanted to regard as meaningful, and excluded everything else, but they failed.

It might seem that this anti-metaphysical movement, initiated by Moore and Russell and developed by the Logical Positivists, would be better able to give centre stage to the HW/PhU problem, in view especially of the central role given to science. But this did not happen, for several reasons. The analytic view of philosophy rendered the HW/PhU problem — a problem concerning the real world — beyond the scope of philosophy. In order to formulate the HW/PhU problem one needs to appeal to metaphysics, the metaphysics of physics, namely physicalism; but Logical Positivism held metaphysics to be meaningless. Again, the central doctrine of Logical Positivism — the verification principle — led to the view that factual scientific statements are about actual and possible sense data; but this amounts to a form of idealism, to the denial of the existence of the physical universe independent of human experience. Once again, the HW/PhU problem cannot even be formulated because the physical universe, that which poses the problem, is removed from view.

Logical Positivism had an immense impact on much subsequent philosophy, especially in the English speaking world,

---

[46] Logical Positivism became well-known in the English speaking world as a result of A.J. Ayer's racy exposition in his *Language, Truth and Logic*: see Ayer (1960).

long after its demise. Somewhat like Cartesian dualism, implications of the doctrine continued to be influential even though the doctrine itself had been rejected. It lent support to the view that philosophy could not be about real problems in the real world — since philosophy is not empirical — and must therefore confine itself to *analysis*, and to producing analytic propositions, as mathematics and logic do.

After the Second World War it was clear that philosophy had split into two mutually hostile camps. On the one hand there is continental philosophy, stemming from the idealist metaphysicians indicated above, against or indifferent to natural science, anti-rationalist, often obscure to the point of incoherence, and including such doctrines as phenomenology, existentialism, critical theory, structuralism, post-structuralism, and postmodernism. On the other hand there is analytic philosophy, stemming from Moore, Russell, Logical Positivism, and Wittgenstein, committed to the idea that the task of philosophy is analysis, lucid about not very much.

Analytic philosophy has never recovered from the disastrous idea that the proper basic task of philosophy is to analyse concepts. This is a recipe for intellectual sterility at best, intellectual dishonesty at worst.[47] Built into the meaning of the kind of words philosophers are interested in — mind, knowledge, consciousness, justice, freedom, explanation, reason, and so on — there are various kinds of often highly problematic *assumptions*; factual, theoretical, metaphysical, evaluative. Instead of imaginatively articulating and critically assessing such assumptions directly, philosophical analysis seeks to arrive at definitive meanings for these concepts as if this can be done in a way which is free of problematic factual and evaluative doctrines. This is a recipe for sterility and dishonesty for, in

---

[47] One of the persistent intellectual sins of philosophy is the idea that philosophical problems need to be solved, can be solved, by an analysis of language, meaning, or concepts. Wittgenstein (1958) is the worst offender. But the idea goes all the way back to Hume, and to Locke.

arriving at such definitive meanings, problematic factual and evaluative doctrines are implicitly decided, but without explicit discussion of these doctrines, and without consideration and critical assessment of alternatives. The whole process is, in other words, profoundly irrational. The classic example of all this is Gilbert Ryle's *Concept of Mind*, which claims merely to analyse the meaning of mental concepts but which thereby, implicitly, espouses behaviourism even though this is explicitly denied.[48]

It may be objected that analytic philosophy has long moved on from this Rylean conception of its task, and no longer confines itself to conceptual analysis. Maybe so; nevertheless, contemporary philosophy has not repudiated fully its analytic past, and is still crippled by it. As a result, it still engages in "puzzle solving", and fails lamentably to take up its proper task.[49]

Neither wing of philosophy has been able to give centre stage to the HW/PhU problem, let alone the more general version of this problem formulated near the beginning of this essay. Neither wing takes its basic task to be to keep alive awareness of what our most fundamental problems are, what our best attempts are at solving them, and what the relative merits and demerits of these attempts are. Neither wing shows even a glimmering of an awareness that this is what philosophy ought to do. Neither wing makes any attempt whatsoever to get schools and universities to grapple, imaginatively and critically, with fundamental problems in a sustained way, and in a way which interacts with more specialized problem solving. I know of no academic philosophers who strive actively to pursue philosophy in such a way, or even conceive of philosophy such a manner.

Perhaps I overstate things a bit here. Certainly Karl Popper did just what I have said a philosopher should do, and thereby

---

[48] Ryle (1949). For a criticism of Ryle and analytic philosophy along these lines see Maxwell (2010b, pp. 667–9).

[49] Popper has decisively criticized doing philosophy via analysis of concepts: see Popper (1963, ch. 2; 1976, section 7).

made immensely significant contributions to thought, especially in his first four books.[50] Bertrand Russell tackled fundamental problems, especially in some of his later, more popular books. J.J.C. Smart, Thomas Nagel, Daniel Dennett, Peter Singer, David Chalmers, and Tim Maudlin have also sought to contribute to thought about fundamental problems.[51] But even here, what is lacking is any awareness of the urgent need to transform academia so that it comes to tackle global problems—global intellectually, and global in the sense of being about the welfare of the planet and humanity—in a lively, imaginative, and critical way, and in a way which both influences and is influenced by specialized problem solving, so that all four elementary rules of reason may be implemented instead of just one.[52] My forty-year long effort to get this message across to my fellow philosophers has been met with indifference and silence.[53]

Over the centuries, academic philosophy has lost its way. What began so promisingly with René Descartes in the 17th century has dwindled either into anti-rationalist, anti-scientific metaphysical nonsense, or into sterile analytic puzzle-solving,

---

[50] Popper (1959; 1962; 1963; 1969).
[51] See, for example, Smart (1963); Nagel (1989); Dennett (1991); Singer (1995); Chalmers (1996); Maudlin (2010).
[52] Recent Philosophy of Science, and Science and Technological Studies, may seem to be branches of philosophy more engaged with science, and with the view of the universe presented to us by science. But these disciplines suffer from the general malaise of rampant specialization, or *specialism* as I have called it (Maxwell, 1980), and fail in their primary philosophical task to try to ensure that academia keeps alive sustained exploration of global problems. See Chapter Four for critical remarks concerning these specific disciplines, and for proposals as to what they ought to do.
[53] For details of my publications related in one way or another to this theme—seven books and over eighty articles (many available online)— see: http://discovery.ucl.ac.uk/view/people/ANMAX22.date.html; and http://philpapers.org/profile/17092. See also my website: http://www.ucl.ac.uk/from-knowledge-to-wisdom.

as far as the mainstreams of philosophy are concerned, ignoring the few exceptions.

Whitehead once said that modern philosophy is a series of footnotes to Plato. It would be more accurate to say that it is a series of footnotes to Descartes. Cartesian dualism is rejected, but problematic implications of the doctrine dominate subsequent philosophy down to today. As I have tried to show, Descartes led to Locke, Berkeley, Hume, Kant, and to the anti-rationalist, anti-scientific nonsense of continental philosophy. The reaction against this led to the esoteric emptiness of analytic puzzle-solving.[54]

Instead of living in Descartes' shadow — and in the shadow of a long series of intellectual blunders made over the centuries — what we philosophers need to do is to return to the fundamental problem which Descartes tried, and failed, to solve — the problem I articulated at the outset: *how can our human world, and the world of sentient life more generally, imbued as it is with the experiential, consciousness, free will, meaning, and value, exist and best flourish embedded as it is in the physical universe?*

---

[54] There is an historical story that analytic philosophers sometimes tell to excuse the poverty of their discipline. It goes like this. "Once upon a time, philosophy all but encompassed the whole of rational inquiry. Then, after Newton, natural philosophy broke away and became independent natural science. Then, after the 18th-century Enlightenment, the social sciences became independent of philosophy. And then logic and linguistics established themselves as independent disciplines. Little was left for philosophy to do apart from conceptual analysis." But this story ignores that there are fundamental, problematic metaphysical, value, and political assumptions inherent in the aims of science which require sustained imaginative and critical — that is, rational — exploration by philosophy. It ignores that fundamental problems, spanning specialized disciplines, need sustained rational attention — a task for philosophy, whether done by academic philosophers, scientists, or others. Far from intellectual developments since the 17th century demanding that philosophy restrict its scope, it is all the other way round. Rampant specialization and fragmentation of research makes the task of engaging in fundamental problems all the more important and urgent.

6. Remarks on How to Solve the Human World/Physical Universe Problem

What is the solution to the problem, granted that Cartesian dualism is untenable? Elsewhere I have written extensively on the subject,[55] so here I confine myself to a few brief remarks.

The person who has made a greater contribution towards solving the problem than any other is not a philosopher at all. He is a scientist: Charles Darwin (1809–1882). Darwinism helps explain how and why purposeful living things can evolve — have evolved — in a physicalistic universe. We need, however, to adopt a version of Darwinism which recognizes that the mechanisms of evolution themselves evolve as life evolves, purposive action playing an increasingly important role, especially when evolution by cultural means comes into play as a result of learning and imitation.[56] We human beings are, above all, the products of evolution by cultural means. Such a version of Darwinism enables us to see that Darwinian evolution merges seamlessly with human history.

Cartesian dualism blunders right from the outset, when it assumes that physics could be in principle comprehensive and complete about the world around us. Actually, physics, and that part of science in principle reducible to physics, seeks to depict only a highly selected *aspect* of all that there is — the causally efficacious aspect, as it might be called, that aspect which determines how events unfold. Theoretical physics seeks to depict that which everything has in common with everything else, and that which needs to be specified in order that a description of a state of affairs at one instant can imply descriptions of states of affairs at subsequent instants — descriptions couched in exactly the same terms.[57] This does not mean that a complete physics would tell us everything factual about the world around us. It

---

[55] See works referred to in note 3 on page 13.
[56] For an account of Darwinian evolution along these lines see Maxwell (2010a, ch. 8). See also Maxwell (2001, ch. 7).
[57] Maxwell (1968a; 1998, pp. 141–55).

would not necessarily tell us about what things look like, sound like, feel like, or what it is like to *be* a certain kind of physical system (a living person). Colours, sounds, tactile qualities will be ignored by physics if they play no role in the predictive and explanatory task of physics.

An elementary argument establishes that physics cannot predict and explain experiential qualities. We can only know what redness as we see it is if we have, at some stage in our life, experienced the visual sensation of redness. A person blind from birth does not know what redness is. Such a person is not, however, debarred thereby from understanding all physics, including optics and the theory of colour perception. He or she is not debarred from understanding all implications of physics. That means physics cannot predict something like "This rose is red" (where "red" is to be understood as the colour we see), however complete the physical description of the rose and its environment may be.[58]

But this built-in incompleteness in principle of physics does not matter. Redness, and other such experiential qualities, do not need to be depicted for physics to perform successfully its predictive and explanatory tasks.

Furthermore, physics must omit these experiential qualities. If it included them, the beautifully unified, explanatory theories of physics would become horrendously disunified and non-explanatory, because endlessly many complex postulates linking physical conditions and experiential qualities would be added to physical theory. This would destroy the unity and explanatory power of physical theory. Thus omitting the experiential is the price that physics pays to be able to develop the marvellously unified and explanatory theories it has developed.[59]

---

[58] This argument is usually attributed to Nagel (1974) and Jackson (1986), although I first spelled out the argument some years earlier in Maxwell (1966; 1968a,b): see especially Maxwell (1966, pp. 303–8; and 1968b, pp. 127 and 134–7). For more on this, see Chapter Five (page 159).

[59] Maxwell (2000b; 2010a, ch. 3; 2011a).

What all this means is that the silence of physics about colours and sounds as we experience them provides no grounds whatsoever for holding that they do not exist out there in the world. Physics is designed by us specifically to avoid any mention of such qualities or properties. We should take it that the world is, in part, as we experience it to be (except when we are suffering from illusions or hallucinations). We see what we ordinarily take ourselves to see, aspects of the world around us.[60] This is, of course, just what Darwinian evolution would arrange for us to be able to see. Animals which could not see aspects of their environment but only the contents of their own minds (as Cartesian dualism would have it) would not last long in the real world.

This view has major implications for the mind/brain problem. It implies that this problem is analogous to, for example, the green grass/molecular structure problem. That brain processes can be conscious sensations, feelings, thoughts, ought perhaps to be no more mysterious than that a leaf (with a certain molecular structure) can be green.[61]

Descartes, Locke, Berkeley, Hume, Kant, and a majority of philosophers who followed made a disastrous mistake in accepting what Cartesian dualism would seem to imply: we really, most directly see not aspects of things in the environment around us but rather the contents of our minds. This blunder, perhaps more than any other, has condemned so much philosophy to foolishness, irrelevance, and triviality.[62]

---

[60] See especially Maxwell (2010a, ch. 3). See also Maxwell (1966; 1968a,b; 1984, ch. 10; 2000b; 2001, ch. 5; 2011a).

[61] See Maxwell (2010a, pp. 77–82). See also Maxwell (2000b and 2011a).

[62] Some ordinary language philosophers have argued that what we see most immediately and directly are not sense data but rather objects in the world around us: see Austin (1962) and Ryle (1949). These philosophers base their arguments on an appeal to ordinary language—not a convincing way to establish the point. And, being constrained by the crippling straitjacket of conceptual analysis, these philosophers failed to return philosophy to the HW/PhU problem—and were entirely incapable of doing that.

7. Conclusion

The world is heading towards disaster. If we continue as we are, climate change and the growth in the world's population may result in millions, possibly billions, dying from starvation, floods, fire, and war. In order to avoid these impending disasters, or cope with them as best we can to the extent that we do not, we need to learn how to do it. And for that in turn we need our institutions of learning—our schools and universities—to be rationally designed, well designed, to help us learn what to do and how to do it. That means, as a bare minimum, that universities give intellectual priority to the tasks of sustaining and promoting imaginative and critical exploration of our most urgent, fundamental problems of living—including global problems—and what we need to do about them. At present universities do nothing of the kind. From the past we have inherited the view that universities must, as a first priority, seek knowledge. First, knowledge is to be acquired; then, secondarily, it can be applied to help solve social problems. This excludes problems of living from the intellectual domain of inquiry, or pushes them to the periphery, and marginalizes them. As a result, universities fail to do what they most need to do to help humanity learn how to make progress towards as wise, as good, a world as possible. As we have seen, *three* of the four most elementary rules of reason are violated in a structural fashion. Universities devoted to the pursuit of knowledge in a way which is dissociated from a more fundamental concern with problems of living betray reason and, as a result of this betrayal, betray humanity.

This is a philosophical disaster. It is the result of the failure of philosophy to establish, within academia, that sustained attention needs to be given to fundamental problems, both intellectual and practical. It is the result of the failure of academic philosophers even to entertain the idea that it is a basic task of philosophy to ensure that universities keep alive imaginative and critical—i.e. rational—thinking about our fundamental problems. For far too long philosophy has monumentally misunderstood what its basic task ought to be.

We need a revolution in academic philosophy so that we philosophers come to do all that we can to bring about a revolution in academia so that our fellow academics in turn come to do all that they can to help bring about a revolution in the world so that humanity begins to tackle in increasingly effective and cooperatively rational ways the immense global conflicts and problems we face.

Chapter Three

How Can Our Human World Exist and Best Flourish Embedded in the Physical Universe?

A Letter to an Applicant to a New Liberal Studies Course

Introduction

In this chapter I sketch a liberal studies course designed to explore our fundamental problem of thought and life: How can our human world exist and best flourish embedded as it is in the physical universe? The fundamental character of this problem provides one with the opportunity to explore a wide range of issues. What does physics tell us about the universe and ourselves? How do we account for everything physics leaves out? How can living brains be conscious? If everything occurs in accordance with physical law, what becomes of free will? How does Darwin's theory of evolution contribute to the solution to the fundamental problem? What is the history of thought about this problem? What is of most value associated with human life? What kind of civilized world should we seek to help create? Why is the fundamental problem not a part of standard education in schools and universities? What are the most serious global problems confronting humanity? Can humanity learn to make progress towards as good a world as possible? These are some of the questions that can be tackled as an integral part of exploring the fundamental problem. But the course does not merely wander at random from one issue to another. Taking the fundamental problem as central provides the course with a coherent structure. The course would be conducted as a seminar, and it would respond to queries and suggestions from students.

Thank you so much for your query concerning our new Liberal Studies Course. I will do what I can to tell you about the Course. It has been in the planning stage for some time. Now at last it will begin, for the first time ever, in the autumn. Those of us involved in creating the Course are very excited about it. We are full of enthusiasm, and we hope our students will be as well.

Our basic idea is that the whole Course should be organized around the exploration of an open, unsolved, fundamental problem. Instead of providing answers to questions never stated or asked (as is so often the case in education), we will together, students and staff, explore imaginatively and critically, that is *rationally*, a real, unsolved, fundamental problem.

The problem we have chosen can be stated quite simply like this:

Fundamental Problem: How can our human world—and the world of sentient life more generally—imbued with the experiential, consciousness, free will, meaning, and value—exist and best flourish embedded as it is in the physical universe?

We interpret this fundamental problem in such a way that it encompasses all of academic thought, from theoretical physics, mathematics, and cosmology, via the biological and technological sciences, to social inquiry and the humanities. It also encompasses literature, music, and the other arts, politics, law, journalism, industry, agriculture, and finance, and indeed all practical problems of living—problems facing individuals, groups, institutions, societies, nations, and humanity as a whole. It is, in our view, quite simply, our fundamental problem—our fundamental intellectual problem of knowledge and understanding, and our fundamental practical problem of living faced by each one of us personally in life, and faced by all of us together.[1] A part of what the Course will attempt to do is see how this, our fundamental problem, connects up with more specific problems—problems of science, of social inquiry and the humanities, political and economic problems, problems each one of us face individually in life as we live—and problems that face humanity as a whole. We will try to trace out a kind of intellectual architecture of problems—the great nave of the intellectual cathedral breaking up into arches, chapels, diverse crooks and crannies of specialized research. And of course we will explore rival ideas as to what our fundamental problem is, how it should be formulated. And we will consider the merits and demerits of these rival ideas.

When we are young we endlessly ask questions. Why, why, why, we demand. Why is that dog barking? Why is the sky blue? What is electricity? Where did yesterday go? What is that bird doing? What are dreams? Why must we die? Then, education gets a hold of us, sits us down, tells us to shut up and listen. "We ask the questions", education says. "Or rather, we don't. None of us here ask questions—except to elicit responses you are supposed to have learned. Only very special people ask questions: research scientists. What we do here is to give you answers. We don't tell you what the questions were that led to these answers. We certainly don't allow you to think about the questions before telling you the answers. And we don't allow

---

[1] For sustained exploration of this problem see Maxwell (2001; 2010a). For the argument that academic inquiry needs to sustain exploration of fundamental problems in order to be rigorous, see Maxwell (1980). Most of my articles and Maxwell (2010a) are available online at http://discovery.ucl.ac.uk/view/people/ANMAX22.date.html. If not available there, try http://philpapers.org/profile/17092.

you to ask serious questions about what you are learning, and why you are learning it. Only when you have gone through a very long process of learning up an awful lot of answers may you, if you are very clever and persistent, be asked to follow up a very specialized question that we, the providers of education, have determined for your PhD."

Of course no teacher or lecturer at school or university ever says any such thing. It is just implicit in much that goes on at school and university. But doctrines never spoken that are implicit in everything that is spoken are all the more powerfully indoctrinating, just because the victim has no idea he is being indoctrinated and so is unable to criticize and reject. It is, in any case, this doctrine, this educational practice of imparting answers without exploration of questions, this tyranny of answers-without-questions, that we intend to break with. All the emphasis will be on the rational tackling of our fundamental problem.

As Einstein once said, "It is, in fact, nothing short of a miracle that the modern methods of instruction have not yet entirely strangled the holy curiosity of inquiry; for this delicate little plant, aside from stimulation, stands mainly in need of freedom; without this it goes to wreck and ruin without fail".[2] Consider what very young children manage to learn, without any kind of formal instruction, when given their freedom to do so, with plenty of stimulation. A child of three has, for herself, learned to understand a language and speak; she has created for herself a whole view of the world and a philosophy of life, a cosmology and some understanding of the social world she finds herself in. No wonder very young children are interested in philosophy: it is for them, as it is not so much for adults, a necessity, a matter of everyday practical learning and discovery. Furthermore, our three-year-old has learned how to act in the world, do things, manipulate spoons, dolls, cups, and other utensils; and she has learned how to communicate and do

---

[2] Einstein (1949, p. 17).

things with others. And all this without a whisper of formal instruction of the kind she will soon receive in school anywhere to be seen. Compared with these mighty intellectual achievements, what Einstein did in creating relativity theory, or Darwin did in developing his theory of evolution, seem meagre indeed.

We then take these infant geniuses, these mighty intellectuals, these powerhouses of curiosity, discovery, and imaginative invention, and we sit them down and tell them to shut up and listen to us. We patronize and humiliate them to an extent that is beyond belief. No wonder so many children become stupefied with boredom at school. What a miracle, indeed, it is that some fragments of the holy, delicate little plant of curiosity somehow survive in some of us, despite a decade or so of educational efforts to stamp it out.[3]

Our Liberal Studies Course swims against the tide. We do everything we can to fan the flames of curiosity into a roaring hunger to seek out, explore, discover, find out, and wonder — to thoroughly mix our metaphors. We do not crush curiosity with endless unasked-for answers. We stimulate it with vistas of questions and problems — a whole intellectual cathedral of mystery. The astonishing wonder of this strange world is all around us, and within us too, as will become all too apparent as the Course proceeds.

You are not a three-year-old, of course. You are not an infant genius of bafflement. But you are closer to that than we are — your teachers. So we hope you will do much to contribute to the fierce questioning and explorations of the Course.

"But does it make any kind of sense", you may ask, "to expect undergraduates to survey and acquire an understanding of all of natural and technological science, all of social inquiry and the humanities, all of our global problems, in a mere three-year undergraduate course? Only in Renaissance times was it possible for a very few mature geniuses to get a grasp of the

---

[3] For a proposal as to how education for young children might proceed so as to capitalize on childish questioning, see Chapter One.

entire culture. Nowadays no one can do it. And yet we poor undergraduates, doing your Course, are expected to acquire such an overview of our whole vast, intricate, endlessly specialized culture in a mere three years. You are asking us to do the impossible!"

It is just this attitude that we had to combat in seeking to set up our Liberal Studies Course in the first place. All too many orthodox, authoritative figures in our university felt as you feel: either the Course would be attempting the impossible; or, given that it would fail to do that, it would be a shallow, pretentious washout, providing nothing more than a simulacrum of real, serious education.

Each one of us, whoever we are, spends our life in a state of supreme ignorance. Nevertheless we live in this world we only very partially and imperfectly know and understand. What really matters is that we have a sufficiently good, rough grasp of the whole to find our way around, aware of our ignorance but also able to learn what we need to learn, as and when it is needed or desired. We need a sympathetic interest in the diverse worthwhile endeavours of humanity, a sense of what theoretical physics at its best seeks to discover, an awareness of what great art or music achieves, some understanding of the good and the havoc that doctors, industrialists, politicians, and bankers can do in the world. And in order to acquire these things, it is essential that we ourselves grapple, in a serious and sustained way, with our fundamental problems, the global problems that lie at the root of science, art, politics, and life. If one has oneself thought long and hard about what kind of universe this may be then, however inadequate this thinking may have been, Newton, Faraday, Darwin, Einstein, and all those other scientists become colleagues, friends, and allies, assistants in one's own efforts to work things out, rather than authorities whose words must be accepted as final.

And besides, science is often made much more difficult than it need be—very much the view of Einstein, incidentally. Physicists often argue that in order to understand physical theory it is essential to have the relevant mathematics at one's fingertips. Without it, all that will be available to one are more or less inadequate metaphors and analogies. But this is quite simply false—and perhaps reflects the inadequate understanding that some physicists have of their own discipline. One almost certainly needs to be a skilled mathematician in order to be able to derive empirical consequences from a physical theory —and of course if that is one's idea of what it is to understand a physical theory, then being an expert mathematician is indeed essential. But all that does is to reveal a pitifully inadequate grasp of what it is to understand a physical theory. The essential thing is to understand what it is that the theory asserts about the world *at the theoretical level* and not just at the empirical level.[4]

Take Einstein's theory of general relativity. Predicting phenomena from the theory is horrendously difficult. Nevertheless, even though one has not the faintest idea as to how that is to be done, one can still have an idea as to what the theory asserts about the world at the theoretical level. General relativity transforms gravity into an aspect of space-time. Space-time is four dimensional, three of space and one of time. Gravity, according to general relativity, is nothing more than the curvature of space-time. Matter, or energy density more generally, tells space-time to curve; the curvature of space-time tells matter what path to travel along. Material objects (free of all forces except that of gravity) travel along geodesics—the nearest thing to straight lines in curved space-time. On the earth's two-dimensional curved surface, geodesics are great circles. What does it mean to say of a three- or four-dimensional space that it

---

[4] Anyone inclined to think that knowing something about science has nothing to do with a liberal education should immediately find a copy of Karl Popper's masterpiece, *The Open Society and Its Enemies*, vol. 2 (Popper, 1962), and look up footnote 6 to chapter 11, pp. 283–4, where Popper argues passionately that science is vital to an authentic liberal education. He says, "For science is not merely a collection of facts about electricity, etc.; it is one of the most important spiritual movements of our day. Anybody who does not attempt to acquire an understanding of this movement cuts himself off from the most remarkable development in the history of human affairs".

is curved? How do we tell whether it is curved or not? We draw a sufficiently large triangle in the space, and measure the angles of the triangle. If they equal 180°, the space is flat or Euclidean. If they add up to more than 180°, the curvature is positive, like that of the surface of a sphere (in two dimensions). If they add up to less than 180º, the space has negative curvature, like that of a saddle (in two dimensions). Actual physical space-time, according to general relativity, has variable curvature depending on how much mass, or energy density, is in the vicinity. There are even waves of variable curvature which travel through space at the speed of light.

Not a hint of mathematics, and yet we have before us the elements of what it is that general relativity asserts about the world. Mathematics is needed, of course, to extract empirical predictions from the theory, concerning such things as the orbits of the planets around the sun, the motions of double stars around each other, the formation and character of black holes.

Take any discipline: theoretical physics, anthropology, cosmology, evolutionary biology, history, geology, economics, mathematics, philosophy, genetics, English literature, organic chemistry, politics, linguistics, international affairs, neuroscience, human geography. What matters is that you come to have some sympathetic understanding of what the basic problems and tasks of these disciplines are, some appreciation of the best of what has been achieved in each case, and what still needs to be done. We would hope that you will become aware of some of the technical difficulties confronting each discipline, so that you acquire some appreciation of what you do know and understand, and what you don't, knowledge and understanding fading into ignorance and incomprehension. We would hope you would come to appreciate how each of these disciplines contributes to illuminating our understanding of the fundamental problem to which the Liberal Studies Course is devoted. We would hope, too, that the special vantage point of the Course would enable you to discern inadequacies in current specialist research, and would enable you eventually to put forward proposals as to how such specialist research can be improved so as to make more fruitful contributions to the fundamental problem. But all this would come in a secondary way, incidentally as it were, to the main task: to explore, with laughter and passion, with imagination and scepticism, our fundamental problem.

One immense advantage in giving priority to tackling our fundamental problem is that, as a result, the whole Course has a focus, a structure, a unity, a natural coherence. Take the fundamental problem away, and one is left with a disjointed jumble of specialized disciplines, one damned item after another, intricate answers to questions never posed, let alone explored. Not only is such a diet indigestible. Devoid of the *problems* that gave rise to these disciplines, these specialized answers, we are rendered incapable of assessing their adequacy, their rationality. Education degenerates into indoctrination.

At this point you may ask: "But this fundamental problem that the Course is supposed to tackle—the problem of how our human world can exist and flourish embedded in the physical universe—why is it such a problem? What exactly is the problem?"

Imagine you go for a walk in the countryside with a friend. It is early summer. The sun is out, there are flowers in the meadows, a gentle breeze rustles the leaves of the trees overhead, white puffs of cloud drift across the blue sky, and you say to your friend, "what a day to be alive!" She says "Yes!", and you stop for a picnic.

Now consider what physics tells us about this scene. Everything you have seen, heard, felt, experienced, thought disappears. You, your friend, the ground, grass, trees, and leaves around you are made up of molecules, in turn made up of atoms, in turn made up of electrons, protons, and neutrons, the latter two in turn made up of quarks and gluons (gluons being particles associated with the so-called strong force that glue quarks together). Sunlight is just electromagnetic waves or, if understood quantum mechanically, billions of particles of various energies called photons. Everything, in short, is just a vast conglomeration of just a very few different kinds of fundamental physical particle interacting with each other in accordance with the precise laws of physics. And this includes you and your friend, your brains, everything you say, experience, feel, and think. When you say "what a day to be alive!" all that happens is that potassium and sodium ions are transmitted in waves down neurons in your brain, leading to muscles being contracted in your chest, tongue, and mouth, causing air to vibrate your vocal chords which in turn cause molecules of the air to dart backwards and forwards in the form of a wave which travels through the air, and which, in turn, causes the ear drums of the ears of your friend to vibrate, in turn causing the little bones of the middle ear to vibrate, provoking in turn vibrations in the inner ear, picked up by tiny hairs which cause neurons to fire, transmitting neuron signals to the brain, which in turn lead to vibrations of the vocal chords and in the air corresponding to "Yes!". The green of the grass, the blue of the sky, the sound of your voices, the smells and tastes of the food of your picnic, your experiences of these things, your thoughts and feelings, your intention to speak, your friend understanding what you say, your mutual decision to stop for the picnic, the meaning and value of your walk in the countryside: all these vanish leaving bleak physics behind, fundamental particles interacting with one another in accordance with precise physical law. Even you and your friend disappear. For you to exist, you must be able to act in the world; you must be able to feel, think, experience, perceive, and initiate action, exercise free will. But in this world of physics there is no room for free will; instead of thoughts, feelings, perceptions, decisions, acts of free will, there are just physical processes going on in your brain causing muscles to contract, your physical body to move around and, on occasions, emit noises — vibrations of molecules in the air.[5]

---

[5] The argument that physics cannot predict or explain the experiential is usually attributed to Nagel (1974) and Jackson (1982; 1986), although I (the real author of this essay) first spelled out the argument some years earlier in Maxwell (1966; 1968a,b), as I pointed out in Chapter Two in note 58 on page 43. For more on this, see Chapter Five, page 159.

This is the heart of the problem. How our human world, the world we experience, everything we hold precious and of value, can survive the corrosive acid of physics.

The problem is very, very old. It goes all the way back to Democritus, over two thousand years ago, one of the first to conceive of the world in purely physical terms. Democritus held that the universe is made up exclusively of indestructible atoms which move through the void. And he declared:

> Colour exists by convention; sweet and sour exist by convention: atoms and the void alone exist in reality.[6]

Two thousand years later, in 1632, Galileo expresses a somewhat similar view:

> [T]hese tastes, odours, colours, etc., so far as their objective existence is concerned, are nothing but mere names for something which resides exclusively in our sensitive body, so that if the perceiving creatures were removed, all of these qualities would be annihilated and abolished from existence.[7]

Galileo goes on to point out that if a feather tickles us we hold that the tickling is in us, not in the feather. In a similar way, colours, sounds, and smells are a kind of tickling in us, and are not objective features of things external to us.

All the great natural philosophers associated with the birth of modern science—Descartes, Locke, Huygens, Newton, and the others—agreed with Galileo. And most scientists today agree as well.

One approach to solving the problem is to adopt the view indicated by Democritus. Reality is such that physics is in principle capable of being correct and complete about everything. In the end, only the physical exists, and anything—such as the experiential—not capable of being even in principle reduced to

---

[6] A slightly different translation is quoted in Guthrie (1978, p. 440).
[7] Galileo, *The Assayer*, quoted in Matthews (1989, pp. 56–7).

physics does not really exist. It is just an illusion.[8] But can we really believe that everything we experience, everything that gives meaning and value to our lives—indeed our very existence as persons—is just an illusion?

Another possibility is that indicated by Galileo, and formulated clearly by Descartes.[9] As far as the material world is concerned, physics is in principle capable of being correct and comprehensive. The material world consists exclusively of the physical. But in addition to the physical universe, there is the universe of conscious minds. Each living human brain has, associated with it, a non-physical mind. But this view, usually called Cartesian dualism, faces horrendous problems. How can the mind and the brain interact? This is required if we are to be able to know anything about the world around us, and if we are to have free will. Can we really believe in the existence of Cartesian minds, distinct from but interacting with brains? And does not this view suffer from the fatal flaw that if everything we experience is in our minds, then it seems impossible that we should ever be able to know anything at all about the physical universe around us?

Another possibility is to deny the existence of the physical universe, declare everything to be in the mind, or made up of human experience, even physics in the end being about no more than actual or possible sense impressions.[10] But can we really believe that this vast and ancient cosmos we seem to have discovered, made up of billions of galaxies, each made up of billions of stars, stretching for billions of light years into the distance, some 14 billion years old at least, is all somehow in our minds? It does not seem credible. What science tells us, surely, is that human life is a minute and very recent phenomenon in this

---

[8] For a modern exposition of the hardline physicalism of Democritus, see Smart (1963).
[9] Descartes (1949).
[10] Berkeley (1957).

vast and ancient universe, mostly utterly alien to our human interests and desires.

Yet another possibility—and the one I favour—is that physics needs to be interpreted as being about only a highly selected aspect of everything: the causally efficacious aspect, as it may be called. The silence of physics about colours, sounds, and smells as we experience them, or our inner experiences of these things, is no reason whatsoever to hold that these things don't really exist, because physics is designed specifically to leave these experiential features out. They are not causally efficacious properties, as mass, or electric charge, are. Even if they exist, physics can leave them out because mention of them is not required to fulfil the predictive tasks of physics. Wavelength of light must be mentioned, but that this light is reddish can be ignored. And furthermore, it must be ignored. If physics sought to include the experiential, it would need to add to physical theory a great number of horribly complex additional postulates linking physical states of affairs with experiential qualities: colours, sounds, and smells as we perceive them, and inner experiences. The beautifully unified and explanatory theories we have in physics—Newtonian theory, classical electrodynamics, quantum theory, general relativity, the standard model—would become horribly complex and disunified, and so non-explanatory. Leaving the experiential out of physics is the price we pay to develop the beautifully explanatory physical theories we have developed. Thus, the silence of physics about the experiential is no grounds whatsoever for supposing that the experiential does not really exist. We should believe in the evidence of our eyes, and hold that grass really is green, sky blue, roses red, out there in the world around us.[11]

Even if this fourth view is, broadly speaking, correct (and it may not be), there is still much that we do not know and understand. There are still immense mysteries before us. We don't

---

[11] For expositions of aspects of this view see Maxwell (1966; 1968a,b; 1984 or 2007a, ch. 10; 2000b; 2001; 2011a; and especially 2010a).

understand how and why the experiential and the physical — or the neurological — are correlated in the way that they are.[12] We don't even know how, or where, consciousness is located in the brain. Some of what goes on in the brain *is* consciousness, but much that goes on may support but is not, itself, consciousness. An absolutely elementary question, then, is simply: where in the brain is consciousness to be located? We do not really know.[13] Nothing could highlight more dramatically our profound ignorance of ourselves — how what happens to the person and to the person's brain are interrelated.

So far I have hinted at some of the intellectual issues the Course will explore — problems that arise in connection with understanding how our human world *exists* embedded in the physical universe. But this is only one aspect of our fundamental problem — and perhaps not the more fundamental aspect. There is also the problem of how what is of value in our human world can best *flourish* embedded as it is in the physical universe. What is of most value in life? And how is it to be realized? What kind of ideal global society should we strive to achieve, in the long term? How do we achieve it? What do we need to do to make progress towards such a practical, ideal global society? How can academia help? What kind of academic inquiry would best help humanity learn how to make progress towards as good and wise a world as possible? These agonizing problems lie at the heart of our Liberal Studies Course.

Humanity is in a state of impending crisis. And the fault lies in part with academia. For two centuries or so, academia has been devoted to the pursuit of knowledge and technological know-how. This has enormously increased our power to act which has, in turn, brought us both all the great benefits of the modern world *and* the crises we now face. Modern science and technology have made possible modern industry and agriculture, the explosive growth of the world's population, modern

---

[12] But see Maxwell (2011a) for a suggestion.
[13] For a conjecture see Maxwell (2001, ch. 8).

armaments and the lethal character of modern warfare, destruction of natural habitats and rapid extinction of species, immense inequalities of wealth and power around the globe, pollution of earth, sea, and air, even the Aids epidemic (Aids being spread by modern travel). Above all, the great success of modern science and technology has made possible our most serious threat of all, the impending disasters of climate change, which promise to intensify all our other global problems (apart, perhaps, from population growth which may fall off when millions, possibly billions, begin to die). All these global problems have arisen because some of us have acquired unprecedented powers to act, via science and technology, without also acquiring the capacity to act *wisely*. We urgently need a revolution in universities so that the basic intellectual aim becomes not knowledge merely but rather wisdom — wisdom being the capacity to realize what is of value in life, for oneself and others, thus including knowledge and technological know-how, but much else besides. This revolution would put problems of living at the heart of the academic enterprise, the pursuit of knowledge emerging out of, and feeding back into, the fundamental intellectual activity of proposing and critically assessing possible actions, policies, political programmes from the standpoint of their capacity to help solve problems of living. This revolution would affect almost every branch and aspect of academic inquiry.[14]

Einstein once put what is wrong in a nutshell like this: "Perfection of means and confusion of goals seem — in my opinion — to characterize our age."[15] All our global problems have arisen because, aided by science and technology, we have been able to pursue goals with immense success that seem, in the short term, highly desirable but which, in the long term, are disastrous. We

---

[14] For a detailed exposition of this thesis and argument, see Maxwell (1984 or 2007a). See also Maxwell (1980; 1992; 2000a; 2003; 2004a; 2007b; 2008a,b; 2009a,b; 2012b; 2013b; 2014a). And see too Chapter Two.

[15] Einstein (1973, p. 337).

urgently need to learn how to improve our aims and methods in life, at personal, social, institutional, and global levels. And for that we need a new conception of rationality—aim-oriented rationality, it has been called[16]—specifically designed to facilitate the improvement of problematic aims (the progressive resolution of problems associated with partly good, partly bad aims). A central task for academia, for universities around the world, is to help people learn how to build aim-oriented rationality into personal and institutional life so that problematic aims may be progressively improved. Far from engaging in such a task, academia fails to put aim-oriented rationality into practice itself. It fails even to acknowledge or recognize the highly problematic character of its own aims, and the need to improve them.[17]

Even science fails to do this. The official view of the scientific community is that the basic intellectual aim of science is truth, the basic method being to assess claims to knowledge impartially with respect to evidence. But this is nonsense. In physics, only *unified* or *explanatory* theories are ever accepted—or even considered. A theory, in order to be accepted, has to satisfy two requirements. It must be sufficiently empirically successful. And it must be sufficiently unified or explanatory in character. Endlessly many empirically more successful disunified rivals can always be concocted, but these are never considered for a moment just because they are disunified. What this means is that the whole enterprise of theoretical physics (and therefore of natural science) just accepts, as a basic assumption, that some kind of unified pattern of physical law runs through all phenomena. The universe is, in some way, more or less physically comprehensible. The basic aim is not truth *per se*; it is, rather, unified truth, explanatory truth—truth presupposed to be unified or explanatory. But this real aim is far too problematic for the scientific community to acknowledge—because it involves

---

[16] Maxwell (1984 or 2007a, ch. 5).
[17] See works referred to in note 14 of this chapter.

acknowledging that science makes a big, highly problematic, metaphysical assumption: the universe is physically comprehensible. So it is repressed and denied. But the outcome of that denial is that this highly problematic aim—the highly problematic assumption inherent in the aim—cannot be subjected to sustained imaginative and critical discussion, as an integral part of science, to try to improve it. The aim of science does evolve over the centuries, of course, but not as a result of sustained critical discussion.[18]

And it does not stop there. The aim of discovering *explanatory truth* is just a special case of a more general and perhaps even more problematic aim of discovering *valuable truth*. And that is sought in order that it will be made available to people in life, ideally to enrich life, either via intellectual aspects of science, or its practical, technological aspects. Science, in other words, has social, humanitarian, or political aims. All these aims, laden with metaphysics, values, and politics, are deeply problematic, and need sustained imaginative and critical discussion, by scientists and non-scientists alike, if science is to be sensitively responsive to the real needs of humanity. But because of misrepresentations by the scientific community of the basic aims of science—because of firm allegiance to untenable orthodox views about science which hold that science appeals exclusively to evidence—the problematic aims of science are not subjected to sustained imaginative and critical attention, and do not steadily improve as a result. Science does not itself put aim-oriented rationality into practice. Academia more generally does not itself put aim-oriented rationality into practice. Neither are in a position to help humanity do this in social life. And humanity does not do it either—in politics, industry, agriculture, economic activity, finance, international

---

[18] For detailed expositions of this thesis and argument concerning the aims and methods—the philosophy—of science, see Maxwell (1974; 1984 or 2007a, ch. 9; 2005; 2006; 2008a; 2010a, ch. 5; 2011b; 2012a; 2014b; and especially 1998; 2004a; 2007a, ch. 14; and 2013a).

relations, and so on. And the outcome of this almost universal neglect of the most elementary principles of reason (designed to help us improve problematic aims) is that the world is in the mess that it is in, with every prospect of things becoming very grim indeed in the not too distant future.[19]

So, these are some of the issues we hope to explore too in our new Liberal Studies Course. There is a sense in which we will be doing *philosophy*, if by philosophy one means either (a) exploration of fundamental problems, or (b) exploration of aims and methods of diverse worthwhile but problematic endeavours — exploration of the *philosophy* of these endeavours, in other words. But academic philosophy today, on the whole, neglects scandalously to do either of these things, (a) or (b). Analytic philosophy is lost in ever more specialized, esoteric discussion of technical puzzles. And continental philosophy continues to be turgid bombast. Both are forms of antiphilosophy. There is a sense in which this is what the world suffers from: the absence of intelligent philosophy devoted to sustained rational exploration of fundamental problems, rational exploration of problematic aims, and associated methods, of major worthwhile but problematic social endeavours (with the aim of helping to improve them).[20] It is in this sense of philosophy, vital but at present almost universally neglected and ignored, that our Liberal Studies Course is a Course in Philosophy.

But I have gone on for far too long. My enthusiasm ran away with me. I hope, nevertheless, that you will be sufficiently interested to apply to do the Course.

Yours sincerely,
Dean of Studies, Liberal Studies Course

---

[19] See works referred to in note 14 on page 61.
[20] See Maxwell (2003; 2008b; 2012a; 2013b). And see Chapters Two and Four of the present volume. It is in my books that I have put this "aims and methods improving" conception of philosophy into practice in the most detailed way: see, in particular, Maxwell (1984 or 2007a; 1998; 2004a; 2010a; and 2014a).

Chapter Four

What's Wrong with Science and Technology Studies?

What Needs to Be Done to Put It Right?

Introduction

After a sketch of the optimism and high aspirations of History and Philosophy of Science when I first joined the field in the mid-1960s, I go on to describe the disastrous impact of "the strong programme" and social constructivism in history and sociology of science. Despite Alan Sokal's brilliant spoof article, and the "science wars" that flared up partly as a result, the whole field of Science and Technology Studies (STS) is still adversely affected by social constructivist ideas. I then go on to spell out how in my view STS ought to develop. It is, to begin with, vitally important to recognize the profoundly problematic character of the aims of science. There are substantial, influential, and highly problematic metaphysical, value, and political assumptions built into these aims. Once this is appreciated, it becomes clear that we need a new kind of science which subjects problematic aims—problematic assumptions inherent in these aims—to sustained imaginative and critical scrutiny as an integral part of science itself. This needs to be done in an attempt to improve the aims and methods of science as science proceeds. The upshot is that science, STS, and the relationship between the two are all transformed. STS becomes an integral part of science itself. And becomes a part of an urgently needed campaign to transform universities so that they become devoted to helping humanity create a wiser world.

1. High Aspirations of History and Philosophy of Science in the 1960s

I came to *Science and Technological Studies* (STS) by means of a rather circuitous route, via a passionate childhood desire to understand the nature of the universe which, after reading Eddington, transformed into an obsession with mathematics which in turn, when adolescence struck, transformed into a desire to understand people via the novel—all of which I failed at dismally.[1] I then took up the study of philosophy in the early 60s at Manchester University. As a part of the undergraduate course, I was introduced to Oxford philosophy, which appalled me. It struck me as a species of anti-philosophy. I concentrated on philosophy of science. Philosophy might not matter, but clearly science does. Then, in the summer of 1961, I had a revelation: philosophy ought to be not about the meaning of words but about how to live! The profound mystery is not even "What is the ultimate nature of the universe?" but rather "What is ultimately of value in life and how is it to be realized?" The problem with academic philosophy is that it is produced by academic philosophers who have already decided how to live, and have thereby lost all interest in real philosophy, which concerns what to do with our agonizingly brief time alive. I decided to do an MA at Manchester, say what needed to be said, and then escape from the madhouse of academic philosophy.[2]

---

[1] I have more to say about this in the next chapter.
[2] I give a more detailed account of all this in the next chapter. See also my (2009a).

And then I discovered the works of Karl Popper, and I became an occasional student at the LSE. Attending Popper's seminars, I was both immensely impressed and somewhat alarmed.[3] Here at last was a philosopher passionately concerned with profound, real problems of the real world which he tackled with fierce intellectual integrity and great originality. There was first his transformation of science—or at least his transformation of our conception of science. Laws and theories cannot be verified in science, but they can be empirically falsified, and that is how science makes progress. As a result of subjecting theories to fierce sustained attempted empirical refutation, we eventually discover where they go wrong, and are thus provoked into thinking up theories which do even better, until they are in turn refuted. Scientific knowledge is simply made up of our best, boldest imaginative guesses that have survived all our most ruthless attempts at empirical refutation.[4]

Then there was his generalization of this falsificationist conception of science to form a radically new conception of rationality. To be rational is to be critical. Just as science makes progress through subjecting our best conjectures to fierce attempted falsification, so more generally, in all areas of human life, we can best hope to make progress by subjecting our best attempts at solving our problems to fierce criticism. Empirical testing in science is just an especially severe form of criticism.[5]

The entire tradition of western philosophy had got it wrong. Scepticism is not the enemy to be vanquished—or to be indulged until it can go no further, thus revealing a bedrock of certainty, as with Descartes, and many empiricists. Quite the

---

[3] Popper could be ferociously critical in his seminars. Rarely did the speaker get past the title before Popper's attack began. Once he reduced a young visiting speaker—now a well-known philosopher of science—to tears.

[4] See Popper (1959; 1963).

[5] "[I]nter-subjective *testing* is merely a very important aspect of inter-subjective *criticism*, or in other words, of the idea of mutual rational control by critical discussion" (Popper, 1959, p. 44, note 1*). Popper refers the reader to his (1969, chs. 23 and 24)—first published in 1945.

contrary, scepticism is our friend, the very soul of reason. It is by means of imagination subjected to sustained, ferocious scepticism that we can learn, and make progress. Science is institutionalized scepticism.

What impressed me most, however, was the application of these ideas to the profound problem of creating civilization or, as Popper called it, "the open society". Rationality is the critical attitude. But this is only really possible in an "open" society, a society, that is, which tolerates a diversity of views, values, and ways of life. In a "closed" society, in which there is just one view of things, one set of values, one way of life, there can be no possibility of criticism, since to criticize A we need, at least as a possibility, some alternative view B. Thus the rational society is the open society — not a society enslaved to some monolithic, dictatorial notion of "reason", but simply a liberal society that tolerates and sustains diversity of views, values, and ways of life, and can, as a result, learn, make progress, and even create and pursue science.[6]

But the move from the closed to the open society has a severe penalty associated with it. We move from certainty to doubt. Living in the open society requires that we shoulder the adult responsibility of living in a state of uncertainty, of doubt. Everything we believe, everything we hold most dear and value — the very meaning and value of our whole way of life — may be wrong or misconceived. Doubt is the price we pay for civilization, for reason, for humanity, and for science. In his masterpiece, *The Open Society and Its Enemies*, Popper calls this essential doubt "the strain of civilization", and he points out that all too many people cannot bear it, and seek to return to the false certainties of the closed society. Even some of our greatest thinkers have sought to do this, and they are the enemies of the open society — above all, for Popper, Plato and Marx.[7]

---

[6] See Popper (1969).
[7] As in note 6.

I breathed a great sigh of relief. Popper had, it seemed, solved the problems that had so tormented me. The anguish of the 20th century—the nightmare of not knowing how to live with only a few measly decades available to try to find out—had been explicated as being due to our new exposure to global society and to history: exposure to a multitude of contradictory beliefs, values, and ways of life which, inevitably, had the effect of throwing into doubt the validity of one's own entire way of life and set of values.

Popper demonstrated, it seemed to me, that it was possible to be an academic philosopher and yet retain one's intellectual integrity.[8] I moved down to London and got a job as lecturer in philosophy of science in the Department of History of Philosophy of Science at University College London. Larry Laudan and Paul Feyerabend were among my departmental colleagues.

It was an exciting time and place to be doing history and philosophy of science (HPS). London felt like the HPS capital of the world. HPS seemed to be a fledgling academic discipline, having associated with it all the excitement, freshness, high aspirations, and optimism of a new discipline. There was the idea that each wing needed the other: history of science would be blind without philosophy of science, which in turn would be empty without history of science. Natural science seemed to be the one great human endeavour that undeniably made progress across generations and centuries. Aside from mathematics, in no other sphere of human endeavour did this happen—not in art, music, literature, politics, or morality. There was technological progress, certainly, and economic progress too, but these were closely linked to, and dependent on, scientific progress. It was the great task of HPS to work out how science did make progress, and what might be learned from scientific progress about how to make progress in other areas of human life: art, literature, law, education, politics, economics, international relations, personal flourishing and fulfilment. Popper had shown the

---

[8] See, too, the next chapter.

way. But he could hardly be the last word on the subject. Popper's philosophy needed to be applied to itself, and subjected to sustained critical scrutiny in an attempt to improve on it. And there were plenty of contending ideas around. There was Thomas Kuhn's *The Structure of Scientific Revolutions*, which in part agreed with Popper in stressing the existence and likelihood of scientific revolutions, but also violently disagreed with Popper in holding that the dogmatic puzzle solving of normal science was an essential and desirable aspect of science as well.[9] Popper, outraged, called normal science "a danger to science and, indeed, to our civilization"[10] (which makes perfect sense, of course, given his viewpoint). Then there was Imre Lakatos's attempted resolution of Popper and Kuhn in his "Methodology of Scientific Research Programmes", which acknowledged that research programmes have a "hard core" (Kuhn's "paradigm" under another name), and legitimately get pursued with a degree of dogmatism.[11] And there was Paul Feyerabend, who went one further than Popper and argued, in effect, that the plurality of views of the open society would need to be imported into science itself. Severe testing—essential, according to Popper, for empirical scrutiny of theories—requires at least the germ of an alternative theoretical idea. We need actively to develop alternative theories simply to be in a position to test severely the reigning, accepted theory—almost exactly the opposite of what goes on, according to Kuhn, during a period of normal science.[12]

2. Beginnings of the Decline of HPS

I am now going to tell the tale of the sad decline of HPS into confusion, irrationality, and irrelevance. But before I do so, I want to stress that good work has been done and continues to

---

[9] Kuhn (1962).
[10] Popper (1970, p. 53).
[11] Lakatos (1970).
[12] Feyerabend (1965).

be done in both history and philosophy of science despite the fashionable stupidities of both disciplines.[13] My complaint is that those who study science and technology—philosophers, historians, sociologists, and others—could have done so much better during the period under consideration, the mid-1960s up to 2013. Much energy has been expended on idiotic disputes—and urgent and fundamental problems, of great importance for science, and for humanity, have been ignored. HPS lost its way.

There are, on the one hand, those sociologists and historians of science—and a few philosophers—who stress the importance of attending to the social dimension of science but, disastrously, abandon such ideas as that science makes progress, acquires authentic knowledge about the world, improves knowledge of fact and truth, and embodies rationality, and puts progress-achieving methods into scientific practice. On the other hand there are some scientists, and some philosophers and historians of science, who defend orthodox conceptions of science against these sociological, anti-rationalist attacks. I must make it very clear, at the outset, that I am critical of both wings of this dispute. The dispute itself—the "science wars" as the dispute came to be called—is the wrong argument to engage in. It is a symptom of the decline in the high aspirations of HPS in the 1960s. It is a distraction from what really needs to be done: to get the scientific community to acknowledge the real, and highly problematic, aims of science which have, inherent in them, highly problematic assumptions concerning metaphysics, values, and politics. It is here that really dramatic and enormously fruitful developments are to be made—as I shall try to indicate towards the end of this chapter. If those who study science had combined with sympathetic scientists to create greater honesty about the problematic aims of science among the scientific community, we might have today a different kind of science, more intellectually rigorous and of greater human value. We

---

[13] To cite just one recent book in the field that I find very impressive: Harper (2011).

might even have a different kind of academic inquiry, rationally devoted to helping humanity create a wiser world. We might even have a different, wiser world—as I will try to explain in what follows. But first I must tell the sad story of decline.

Somewhat arbitrarily, we may begin with a dreadful blunder made by Feyerabend. On Popper's behalf, he assailed the logical empiricists, Hempel, Carnap, and Nagel, for holding that meaning had to be transported up from evidence to theory.[14] No, Feyerabend argued, that was not possible, for observational terms are "theory laden", so that conflicting theories would have conflicting, or at any rate different, observation terms, conflicting or different accounts of observational phenomena. There can be no such thing, Feyerabend argued, as a stable observational language independent of theory (an argument to be found in Kuhn as well). But logical empiricism depends utterly on there being just such a theory-independent observation language. The whole position takes it for granted. Its non-existence destroys logical empiricism completely. Its foundations do not exist! So far, so so good.[15] But

---

[14] Logical positivism held that the meaning of a proposition, or theory, is the method of its verification. The idea was to render scientific theories meaningful, but metaphysics, which cannot be verified empirically, meaningless. This failed for the simple reason that scientific theories cannot be verified. So logical positivism morphed into the very much weaker doctrine of logical empiricism which held that theoretical terms acquire their meaning as a result of being linked to observational terms by means of bridge statements. It was this doctrine that Feyerabend set out to demolish.

[15] Not really very good, of course, for even if observational terms are theory-laden, nevertheless given any two conflicting theories ostensibly about the same, or overlapping, ranges of phenomena, one can always concoct observational terms that are such that the theory presupposed by them is neutral between the two theories: see my (2014b). That this can always be done means that empirical predictions of conflicting theories about overlapping phenomena can always be assessed in terms of these phenomena described by means of terms that presuppose low-level theory that is neutral between the conflicting theories in question. Feyerabend's argument for incommensurability, methodological anarchy, and dadaism collapses completely. I did my best to point this

then Feyerabend made an idiotic mistake. If meaning cannot be transported up, from observation to theory (because a theory-independent observation language does not exist), then meaning must be transported down, from theory to observation terms. But this means in turn, Feyerabend argued, that conflicting theories, with different theoretical terms, must have different observational terms as well, which in turn means that the predictions of the conflicting theories cannot be compared. And so the very basis for Popper's philosophy of science—his falsificationism—collapses.[16] Not just logical empiricism, but falsificationism too must be thrown on the rubbish dump of history. Scientists should follow their instincts, Feyerabend concludes. Anything goes. Methodological anarchy reigns supreme. There is no such thing as the rationality of science. It is irrational. And it is damaged when it attempts to conform to some misguided idea of rationality dreamed up by a philosopher of science.[17]

Feyerabend had an absolutely disastrous influence. He became a sort of approved intellectual court jester. All those who deplored what they perhaps saw as the illegitimate mighty authority of science were entranced by Feyerabend's annihilation of science's claim to be rational and methodological, upon which its mighty authority rested. The emperor had no clothes. Feyerabend had stripped science bare. Or so it seemed to all too many.

HPS began to take an absolutely disastrous turn for the worse. The initial great ambitions and optimism of the fledgling discipline were lost sight of. HPS began to tear itself to pieces in an orgy of stupidity, like a political party thrown out of power,

---

out to Feyerabend in person, but he was having none of it. And nor was Kuhn when I tried to point out that his argument for incommensurability rested on the same fallacy. The problem was solved long ago by Michael Faraday in scientific practice in connection with his work on electrolysis. How extraordinary that, over a century later, two leading philosophers of science could not grasp what Faraday had understood long ago: see Maxwell (2014b).

[16] Feyerabend (1970).
[17] Feyerabend (1975; 1978; 1987).

or a political movement with no hope of ever gaining power. It came in wave after wave of idiocy.

At about the same time as Feyerabend began to drum up support for relativism and unreason, a very different kind of disastrous stupidity was being incubated in Edinburgh. It was called "the strong programme", and its authors were Barry Barnes and David Bloor.[18] They argued that science is social in character, and therefore needs to be studied by sociologists. This means, they held, that there is no such thing as scientific truth, knowledge, rationality, or progress. There is just change of scientific belief, as science goes on its way. Traditionally it has been held that science is rational, its theories being established by evidence, science being entitled to claim it acquires genuine knowledge of factual truth, science thus progressively increasing and improving our knowledge and understanding of the universe. But all this has been shown to be untenable — by Kuhn, Feyerabend, and others. Those philosophers of science who do, absurdly, still claim that science makes progress, is rational, and acquires genuine knowledge of factual truth are unable to say how this is done. The problem of induction remains unsolved. Even Popper, who almost alone does claim to have solved the problem, has not really solved it. So science must be treated as social in character, purely *social* factors determining what is accepted and rejected in science — namely observational and experimental results, laws, and theories. It is the sociologist of science, not the philosopher of science, who can improve knowledge about science, how it proceeds and modifies its beliefs, its "scientific myths" one might say. Truth, fact, knowledge, scientific progress, method, and reason all fly out of the window. These are fantasy ideas of old fashioned philosophy of science, illusory notions that have nothing to do

---

[18] Harry Collins, John Henry, and others were, and still are (at the time of writing), associated with the movement.

with science as it really is, an integral part of society, social through and through.[19]

At about the time that "the strong programme" was being launched on the world, *The British Society for the Philosophy of Science* held its annual conference in Edinburgh, and naturally the Edinburgh School was given its chance to air its ideas. I remember thinking at the time that ideas as foolish as these would never get anywhere. How wrong I was. I also remember wondering why proponents of "the strong programme" had not bothered to read Popper, for in *The Open Society and Its Enemies* Popper anticipated and decisively dealt with and obviated the need for this sociological programme.[20]

Popper makes the point that rationality—critical rationality, that is—is essentially social in character, in that criticism requires diversity of views (as we have seen) and so many people in communication to hold and discuss these diverse views. Furthermore, science is fundamentally social in character too, and owes its rationality, its scientific character, to its social character. Far from the social character of science somehow cancelling the scientific character of science, as proponents of "the strong programme" seemed to believe, it is all the other way round: the scientific character of science actually requires science to be inherently social.

Furthermore, what *methods* are implemented in scientific/social practice may well, quite obviously, have an immense impact on whether science meets with success in improving knowledge about the world. Compare M_1: "accept theories that are empirically refuted, and reject theories that are empirically confirmed" with M_2: "accept the best explanatory theories that are empirically confirmed, and reject theories that are decisively refuted". We would all agree that a community of scientists that puts M_2 into social/scientific practice is more likely to meet with success and improve knowledge than one that puts M_1 into

---

[19] Bloor (1976); Barnes (1977; 1982; 1985); Barnes, Bloor and Henry (1996).
[20] Popper (1969, vol. 2, chs. 23 and 24).

practice. It is, in short, utterly trivially obvious that what methods are implemented in social/scientific practice may well make a profound difference to the intellectual success or failure of science — its success in acquiring knowledge about the world.

What *methods* science puts into practice is a vital part of the whole social structure of science which the sociological study of science cannot possibly ignore if it is to be remotely adequate. Both Popper and Kuhn are very good, in their different ways, in pointing out that what matters are the methods that are implicit in scientific practice.[21]

Construing science to be a social endeavour thus does not abolish the intellectual or rational character of science, and certainly does not do away with crucial questions about what methods science does, and ought to, adopt and implement. It does not mean that science does not acquire genuine factual knowledge and make progress.

Furthermore, science in particular, and our social world more generally, is imbued with values, whether intellectual, moral, legal, or aesthetic, some better than others. It certainly ought to be a part of the professional job of academics to try to discriminate between good and bad intellectual values, and promote the former. Sociologists of science, like scientists themselves, philosophers of science, and all other academics, ought to do what they can, in their professional work, to promote good intellectual values — ones having to do with rationality, validity, the successful pursuit of knowledge of fact and truth — at the very least.

"The strong programme" is a kind of acid which eats all these things away, and leaves science as a value-denuded, knowledge-denuded, truth-and-reason-denuded, empty social practice. But all this arises from elementary and appalling misunderstandings about the nature of our social world in general,

---

[21] See previous note, and Popper (1974, section 32, "the institutional theory of progress", pp. 152–9). See also Kuhn on normal science: Kuhn (1962, chs. III–V).

and that bit of it that is science in particular — a refusal at the outset to see that values and standards, whether intellectual or humanitarian, are essential features of our social world. To exclude all values from the social world *a priori*, as it were, is to adopt something close to a psychopath's vision of things. Ironically, it probably comes from the unconscious adoption of a very crude philosophy of science which says values have no place in science, and hence no place in sociology, or the sociology of science either (I say "ironically" because, according to the proponents of "the strong programme", philosophy of science is a sort of irrelevant fantasy).

It is as if proponents of "the strong programme" had convinced themselves of the correctness of the following argument.

1. Reason, validity, valid scientific methods, truth, fact, knowledge, and scientific progress are all inherently purely *intellectual*.

2. The *intellectual* is not *social* (and no part of the *social* is *intellectual*).

3. But science is wholly and purely *social*.

4. Hence science is wholly free of the *intellectual*. It has nothing to do with reason, validity, valid scientific methods, truth, fact, knowledge, scientific progress.

The argument may be valid, but step 2 is false. The intellectual is wholly social in character. That makes step 4 false as well. As I have said, one cannot begin to do justice to the character of our social world if one refuses, at the outset, to acknowledge that the social is quite essentially imbued with values of all kinds, intellectual, moral, legal, aesthetic — imbued not just with values but with *what is of value*.[22]

3. Social Constructivism and Anti-Whiggism

I have so far concentrated on the damage done to HPS by Feyerabend's methodological anarchism and the blunders of "the strong programme". But damage came from another

---

[22] For value realism see my (1984 or 2007a, ch. 10; 1999; 2001, ch. 2).

source as well: French philosophy, Foucault, Derrida, and others. The upshot was a whole new way of construing science, which may be called "social constructivism". This is the view indicated above that I have attributed to "the strong programme". Scientific knowledge is merely a social construct, having nothing to do with knowledge, truth and falsity, or reason. In studying science and its history, we must entirely forego the idea that science makes progress, and we must refrain from making intellectual or scientific judgments about one theory being "better", "truer", or "more firmly established" than another. In the main sociologists and historians took to social constructivism, while philosophers of science looked on in amazement and horror at the idiocy of it. As a result, HPS broke asunder. The integrated enterprise, bringing together history and philosophy of science, which had started out with such high hopes and aspirations, and which was still alive and kicking when I began my academic career around 1965, was no more.

An even more devastating consequence, perhaps, of the widespread adoption of social constructivism among historians of science was that it annihilated the fundamental problem of the discipline. As I stressed at the beginning of this chapter, science is almost unique among human endeavours in that it makes genuine progress. We know and understand vastly more about the universe, and ourselves as a part of the universe, than was known to Darwin, to Faraday, to Newton, or to Aristotle. The fundamental problem of HPS is: how has scientific progress come about? And for philosophy of science in particular: how is scientific progress possible? What methods have brought it about? What methods give the best hope of progress?

Social constructivism annihilates these fundamental problems. What ought to be the central problem of the history of science just disappears from view. This is perhaps the strongest indication of the intellectual poverty and destructive character of social constructivism.

Where did this idea that science does not make progress come from? In addition to the intellectual blunders that I have already indicated, it came from a blunder about history. The historian Herbert Butterfield wrote a little book against what he called "Whiggish history".[23] This is history that takes for granted that progress, the spread of enlightenment, democracy, and justice are inevitable, and it is the job of the historian to describe this process. An even cruder kind of Whiggish history would have built into it dogmatic assumptions about what does constitute progress, history being written as propaganda to help the process along, or fool the reader into believing that progress in this sense has occurred and is occurring when nothing of the kind is the case.

Whiggish history in these senses is intellectually disreputable. It is, however, utterly absurd to think that this means historians can't ever write histories of any human endeavour whatsoever that does in fact make progress towards some goal, or seeks to make progress towards some goal. That is, clearly, an absurd position to adopt. If there is a human endeavour that makes progress, or seeks to make progress, then it must be possible to write intellectually decent histories of it. It may be very important to do this. Establishing the *a priori* dogma that *any* such history must be Whiggish—that is, based on the assumption that progress is inevitable or, worse, mere propaganda on behalf of the endeavour—just ensures that no intellectually decent history of any progress-achieving endeavour will be written, an appalling impoverishment of what history should be.

Science is one of those rare human endeavours that does make progress across generations and centuries. It is vitally important that good, intellectually responsible histories of this progress-achieving endeavour of science are written. How is

---

[23] Butterfield (1951). Butterfield seems to believe that ideally history would be about everything. He says at one point, "The value of history lies in the richness of its recovery of the concrete life of the past" (1951, p. 68). He ignores that history is always about something specific—power, the Black Death, the potato, or whatever—and may quite legitimately be about something that seeks, and even perhaps achieves, progress.

this to be done so as to avoid Whiggishness? There are a few very obvious points to make.

1. Do not assume progress is inevitable.

2. Do not write propaganda on behalf of science and scientific progress. Praise where praise is deserved, and criticize where criticisms need to be made. Do not conceal deplorable incidents—faking of results, plagiarism, petty disputes about priorities, immoral or criminal behaviour of scientists. Explore controversial issues about science and politics, science and war, science and the arms industry, science funding.

3. Do not just write about scientific success. In order to understand how and why scientific progress occurs it is absolutely essential to take into account the blind alleys, the research projects that led nowhere, the false leads, the ideas that turned out to be unproductive.

4. Do not hesitate to make judgments about how good or bad a piece of scientific research was. Do not assume, however, that scientific work is good if it turns out to be true, successful, or productive, and bad if it turns out to be false, unsuccessful, or unproductive. Do not judge the intellectual merit of scientific work purely in terms of the contribution it ultimately makes to scientific progress. Brilliant scientific work may lead nowhere, and contributions that turn out subsequently to be important may come out of shoddy work, even out of mistakes.

5. In writing about past scientific episodes, try to see things from the actors' points of view so as to understand their problems, aims, ideas, theories, prejudices, standards, methods as they saw them and experienced them. Seek to assess scientific work and contributions in terms of the standard prevalent at the time. But do not shrink from assessing the merit and significance of past work from the standpoint of the best standards and ideals available to us today, in an attempt to assess the significance of past contributions to overall scientific progress— where it is relevant to do this. Do not shrink from criticizing past work from the standpoint of our best current intellectual standards, should it be relevant to do this.

6. Keep in mind that what constitutes *progress* depends on what *aim* is presupposed. There are a range of aims that may be assigned to science, all more or less problematic (see below). Whether science as a whole, or a particular science, makes progress or not during a specific period may depend crucially on what *aim* for science is presupposed. Consider, for example, the aim of science of "improving human knowledge". This may be interpreted as (1) improving knowledge of scientific experts, or (2) improving knowledge of humanity as a whole. A science might make splendid progress given aim (1), but very little progress or none at all given aim (2).

7. Take into account that, insofar as scientific knowledge is conjectural in character, judgments about scientific progress will be conjectural too. Thus the historian's judgments as to whether scientific progress has taken place, what it consists in, and how it was achieved will be conjectural as well, and may be falsified or at least modified as current scientific knowledge is modified. This is of course more likely to happen to history of recent scientific developments than it is to history of scientific developments a century or so ago.

8. Far from it being assumed at the outset by a history of a progress-seeking endeavour, whether scientific or not, that progress occurs (let alone is inevitable), such a history should be open-minded about the matter. Whether progress has been made, of what type, towards what goal, and of what mixture of advance and regression are all questions open for historical research to discover. It might indeed emerge that no progress has been made, or that the opposite has happened and the endeavour has regressed (perhaps this is the case as far as HPS itself is concerned).

9. Make no *a priori* judgments about whether *intellectual* or (non-intellectual) *social* factors influenced some specific piece of

scientific work.[24] Much that scientists do is probably influenced by a complicated mixture of these factors. Thus the decision to work on a specific scientific problem may be influenced by (1) curiosity, (2) availability of funds, (3) the guess that the problem will be easily solved, and will thus enhance career prospects, (4) the hunch that it will turn out to be important to solve from the standpoint of social applications (medical, industrial, etc.), (5) a request from the scientist in charge of a scientific team, (6) the presence in the laboratory of relevant equipment. Are any of these considerations wholly "intellectual" or wholly "social"?

10. Science is a human endeavour different for the historian from other, non-intellectual endeavours — even endeavours that also make progress. In the case of science, what the historian studies, and the discipline of history itself, have some common goals: to improve human knowledge and understanding. This means that in the particular case of science, it may well be legitimate for the historian to write history which seeks to help promote the very thing he is writing about. The historian of science may quite legitimately seek to highlight neglected work from the past that may, if better known, have important implications for the future of the science in question.[25] This kind of science-promoting history can be done in a thoroughly intellectually responsible way even though, if done about other kinds of endeavour, it might well amount to no more than a kind of propaganda for the endeavour in question. Serious history of science of this kind should not, however, degenerate into the simplified, distorted, potted history that scientists tell their

---

[24] Intellectual factors at work in science are social in character. We can thus distinguish two kinds of social factors influencing science: the intellectual, and the non-intellectual.
[25] I have attempted something along these lines in my (2010a, ch. 10, especially pp. 276–89). Following Alister Hardy's lead in his (1965), I call upon some neglected and misrepresented history of evolutionary thought to provide support for an interpretation of Darwinian evolution which gives an increasingly important role to purposive action in evolution, and which holds that the mechanisms of evolution themselves evolve.

students for pedagogic purposes. The all-important point, furthermore, is that history of science does not have to be science-promoting, in this way, as the above points, 1 to 9, indicate.

As long as these and similar strictures are kept in mind and observed, there is no reason whatsoever why histories of science that depict science as making progress should not be done that meet the highest standards of intellectual excellence, there being not the faintest whiff of Whiggishness in any of the bad senses.

It is quite extraordinary that so many historians of science have been unable for decades to draw the distinction between "Whiggish history" in the bad senses, and "history of some endeavour that makes progress" that is intellectually responsible and excellent. It is all the more extraordinary, when one considers that the failure to draw this obvious distinction has meant that, for these historians, the fundamental problem of the history of science, "How and why has scientific progress come about?", has died, and disappeared entirely from view. It is as if cosmologists managed to reach the conviction that there is no such thing as the cosmos, or biologists convinced themselves that there is no such thing as life on earth. Intellectual history is turned into mere social gossip.

I encountered the consequences of these elementary intellectual blunders in my professional life as a lecturer in philosophy of science in the Department of History and Philosophy of Science at University College London. We taught a joint MSc programme with the Wellcome Institute, with Bill Bynum, Chris Lawrence, and Mike Neve—these latter all firmly committed to social constructivism in the history of medicine. Students were baffled. At the Wellcome Institute they learnt there is no such thing as scientific progress, rationality, truth, or knowledge. In my lectures they heard that there is a fundamental problem concerning the rationality of science—a big, serious, unsolved problem about how it is that science acquires knowledge of truth and makes progress. How to choose between holding that truth, knowledge, progress, and reason are of fundamental importance, and holding that there are no such things at all? In the end most shut their eyes and made a Kierkegaardian leap of faith into one or other position. I pleaded with Bynum, Lawrence, and Neve to hold a seminar with me and the students to discuss these issues. They refused. One year I did persuade one historian, Rob Iliffe, to take part in such a discussion of the issues, but only if it was informal, after hours as it were, and with beer to drink. He pointed out how bad it is just to assume dogmatically that science makes progress when there is much to criticize in modern science. I replied that if rationality is abandoned, the very possibility of being critical of modern science is abandoned too, for criticism presupposes and requires rationality. Iliffe had no answer. In the end he was reduced to arguing that he had to go along with social constructivism in order to get an academic job as an historian of science.

One bizarre feature of social constructivism is that its proponents are often left wing and highly critical of aspects of modern science. But of course, as a result of abandoning rationality, the very possibility of criticism disappears. My attempts to point this out to proponents of "anti-Whiggism" over the years invariably fell upon deaf ears.

Sometime in the 1970s and 1980s a new branch of HPS emerged which came to be called *Science and Technology Studies* (STS). This emerged out of the sociology of science, out of a concern to give far greater emphasis to technology and the technological sciences, and out of a concern to tackle issues associated with science and society — the impact of science on society, and *vice versa*. From the outset, much of the potential inherent in STS has, however, been subverted by the influence of ideas stemming from "the strong programme", social constructivism, "anti-Whiggism", and anti-rationalism.

There has been a tendency too for Philosophy of Science to degenerate into a kind of scholasticism in that it has splintered into a multitude of specialized disciplines: philosophies of the specialized sciences — physics, chemistry, neuroscience, astronomy, botany, and so on. As a result, Philosophy of Science has rather lost sight of the magnificent endeavour of natural science as a whole, and has come to ignore the great, fundamental problems that were, initially, the whole *raison d'être* for its existence: the problem of induction, the problem of the rationality of science, the problem of how, by what means, science makes progress.

4. Alan Sokal's Hoax and the Science Wars

In 1996 the worst excesses of the social constructivists and anti-rationalists were brilliantly satirized by a spoof article by Alan Sokal called "Transgressing the Boundaries; Toward a Transformative Hermeneutics of Quantum Gravity".[26] This was published in an American academic journal called *Social Texts*, the editors of which took the paper to be a serious academic contribution. Actually, it was a tissue of hilarious nonsense decked out with liberal quotations from the constructivists — although Sokal admitted subsequently that, despite considerable effort, he did not always succeed in the article in attaining the dense obscurity of what he satirized. One of the editors was interviewed on the BBC, on the *Today Programme*, and made the dreadful mistake of protesting at the immorality of Sokal's hoax instead of laughing and admitting that they had been had.

Around this time, and partly in response to Sokal's hoax, the "science wars" exploded onto the scene, some scientists and philosophers of science springing to the defence of science against the corrosive acid of social constructivism, anti-rationalism, and postmodernism. Paul Gross and Norman Levitt wrote a book assailing the worst excesses of postmodernist writing about science, and subsequently edited a book that continued the argument.[27] Alan Sokal and Jean Bricmont outraged French intellectuals with devastating criticisms of French philosophers' writings about science: Jacques Lacan, Luce Irigaray, Bruno Latour, Gilles Deleuze, and others.[28] Noretta

---

[26] Sokal (1998). See also Sokal (2008) for an annotated version of the hoax article, and essays on related matters.
[27] Gross and Levitt (1994); Gross, Levitt and Lewis (1996).
[28] Sokal and Bricmont (1998).

Koertge edited *A House Built on Sand: Exposing Postmodernist Myths About Science*.[29] Others joined the affray. Social constructivists protested that distinctions were being ignored, contexts overlooked.

Did this counter-attack on behalf of orthodox conceptions of science win the day and rid STS of anti-rationalist views? No. They continued to be influential, but in perhaps a slightly muted way. Here is just one fairly recent example of this influence, and how damaging it can be.

In 2009 a young practitioner of STS, Sergio Sismondo, gave a good lecture on the scandal of medical "ghost writing" in my very department of STS at UCL. "Ghost writing" is the process whereby a drug company writes a paper specifically designed to be published in a particular medical journal, in terms of such spurious features as layout, references, etc. The paper praises a new drug the firm has produced, and then gets an academic who is an acknowledged authority in the field to author the paper, even though he or she has not seen data from trials, in particular data about harmful side effects. The paper is duly published, and what is essentially an advertisement is treated by GPs and other researchers as if it is a genuine contribution to scientific knowledge.

When the talk was over, I made the point, dressed up as a question, that such a contribution could not be regarded as an authentic contribution to knowledge. The deception might well lead to deaths—as happened in connection with Vioxx. No, Sismondo responded, such a paper did constitute a contribution to scientific knowledge because it had satisfied all the criteria for publication of the journal in question—and there could be no question about some of these criteria being epistemologically irrelevant. Social constructivist habits of thought had rendered

---

[29] Koertge (1998).

Sismondo incapable of acknowledging the full extent of the scandal, even the criminality, his talk was about.[30]

Social constructivists and their sympathizers are absolutely right to stress the vital importance of taking social aspects of science and technological research into account. The way this has been done, however, has been an intellectual disaster. It has helped sabotage urgently needed developments in thinking about science which would have brought together scientific and social thinking in sensible, rational, and fruitful ways, as I shall try to show in a moment. The point that there is no agreed solution to the problem of induction, the problem of the rationality of science, is absolutely correct. The solution is, however, waiting in the wings to be taken note of. And this solution leads on to a profound transformation in the way we think about the aims of science, science itself, and academic inquiry as a whole, more generally.

5. Metaphysics, Values, and Politics Inherent in the Aims of Science

Most scientists and philosophers of science take for granted one or other version of a view of science that I have called *standard empiricism* (SE). This holds that the basic intellectual aim of science is factual truth (nothing being presupposed about the truth), the basic method being to assess claims to knowledge impartially with respect to evidence. Considerations such as the simplicity, unity, or explanatory character of a theory may influence what theory is accepted, but not in such a way that the universe or the phenomena are permanently assumed to be simple, unified, or comprehensible. According to SE, what theory is accepted may even be influenced for a time in science by some paradigm or metaphysical "hard core" in the kind of way depicted by Kuhn and Lakatos[31] as long as, in the end, empirical

---

[30] See Sismondo (2009a). For a criticism see McHenry (2009b); and for a reply see Sismondo (2009b).
[31] Kuhn (1962); Lakatos (1970).

success and failure are the decisive factors in determining what theories are accepted and rejected. The decisive tenet of SE is that *no substantial thesis about the nature of the universe can be accepted as a permanent part of scientific knowledge independently of empirical considerations* (let alone in violation of empirical considerations).

Even those who—like Feyerabend, social constructivists, and postmodernists—reject the whole idea that science is rational, delivers authentic knowledge, and makes progress nevertheless tend, in a way, to uphold some version of SE as the only possible rationalist conception of science. No rational account of science is possible, they hold in effect, because the only candidate, SE, is untenable (as shown by the failure of SE to solve the problem of induction).

Despite being almost universally taken for granted by scientists, SE is nevertheless untenable. SE very seriously misrepresents the aims of science. The intellectual aim of science is not to improve knowledge of factual truth, nothing being presupposed about the truth. On the contrary, science cannot proceed without making a very substantial and highly problematic *metaphysical* hypothesis about the nature of the universe: it is such that some kind of unified pattern of physical law governs all natural phenomena. Science seeks not truth *per se* but rather *explanatory* truth—truth presupposed to be explanatory. More generally, science seeks *valuable* truth—truth that is of intrinsic interest in some way or useful. This aim is, if anything, even more problematic. And science seeks knowledge of valuable truth so that it can be used in social life, ideally so as to enhance the quality of human life. There are, in other words, problematic *humanitarian* or *political* assumptions inherent in the aims of science. In holding that the basic intellectual aim of science is *truth per se*, the orthodox position of SE misrepresents the real and highly problematic aims of science.

The vital task that needs to be done to develop STS in fruitful directions—a task not performed because of the influential absurdities of "the strong programme", social constructivism, and the science wars debate—is to give absolute priority to two fundamental questions: What are the real aims of science? What ought they to be? Ever since around 1970, when I began to consider these questions, those associated with HPS and STS ought to have put these two questions at the heart of science studies. If this had been done, science studies, in conjunction with sympathetic scientists, science journalists, and others, might have helped develop a conception of science, and even a kind of science, both more rigorous and of greater human value than what we have today. Indeed, a new kind of academic inquiry might have emerged that is rationally devoted to helping humanity make social progress towards as good a world as possible. We might even have begun to see the beginning of a new kind of world capable of tackling its immense global problems in increasingly effective and cooperatively rational ways. None of this has come about because the academic disciplines most directly responsible for helping to initiate these developments, HPS and STS, have been distracted by intellectual stupidities.

The key step that needs to be taken to permit these urgently needed intellectual, institutional, and humanitarian developments to unfold is the widespread recognition that standard empiricism (SE) is indeed untenable, and needs to be replaced by something better. So, let us see why SE is untenable.

As it happens, reasons for rejecting SE have been spelled out in the literature again and again, ever since 1974.[32] But these refutations of SE have been ignored. In outline, the refutation goes like this.

Theoretical physics persistently only ever accepts *unified* theories—theories that attribute the same dynamical laws to the phenomena to which the theory applies. Given any such accepted theory—Newtonian theory, classical electrodynamics, quantum theory, general relativity, quantum electrodynamics, or the standard model—endlessly many disunified rivals can be

---

[32] See my (1974; 1993a; 1998; 2000c; 2002; 2004a; 2005; 2007a, chs. 9 and 14; 2009a,c; 2011b; 2013a).

easily concocted to fit the available phenomena even better that the accepted unified theory.[33] These disunified rivals that postu-

---

[33] Here is a demonstration of this point. Let T be any accepted fundamental physical theory. There are, to begin with, infinitely many disunified rivals to T that are *just as empirically successful* as T. In order to concoct such a rival, T_1 say, all we need to do is modify T in an entirely *ad hoc* way for phenomena that occur after some future date. Thus, if T is Newtonian theory (NT), NT_1 might assert: everything occurs as NT predicts until the first moment of 2050 (GMT) when an inverse cube law of gravitation comes into operation: $F = Gm_1m_2/d^3$. Infinitely many such disunified rivals can be concocted by choosing infinitely many different future times for an abrupt, arbitrary change of law. These theories will no doubt be refuted as each date falls due, but infinitely many will remain unrefuted. We can also concoct endlessly many disunified rivals to T by modifying the predictions of T for just one kind of system that we have never observed. Thus, if T is, as before, NT, then NT_2 might assert: everything occurs as NT predicts except for any system of pure gold spheres, each of mass greater than 1,000 tons, moving in a vacuum, centres no more than 1,000 miles apart, when Newton's law becomes $F = Gm_1m_2/d^4$. Yet again, we may concoct further endlessly many equally empirically successful disunified rivals to T by taking any standard experiment that corroborates T and modifying it in some trivial, irrelevant fashion—painting the apparatus purple, for example, or sprinkling diamond dust in a circle around the apparatus. We then modify T in an *ad hoc* way so that the modified theory, T_3 say, agrees with T for all phenomena except for the trivially modified experiment. For this experiment, not yet performed, T_3 predicts—whatever we choose. We may choose endlessly many different outcomes, thus creating endlessly many different modifications of T associated with this one trivially modified experiment. On top of that, we can, of course, trivially modify endlessly many further experiments, each of which generates endlessly many further disunified rivals to T. Each of these equally empirically successful, disunified rivals to $T-T_1, T_2,... T_\infty$—can now be modified further, so that each becomes *empirically more successful* than T. Any accepted fundamental physical theory is almost bound to face some empirical difficulties, and is thus, on the face of it, refuted—by phenomena A. There will be phenomena, B, which come within the scope of the theory but which cannot be predicted because the equations of the theory cannot (as yet) be solved. And there will be other phenomena, C, that fall outside the scope of the theory altogether. We can now take any one of the disunified rivals to T, T_1 say, and modify it further so that the new theory, T_1^*, differs further from T in predicting, in an entirely *ad hoc* way, that phenomena A, B, and C occur in accordance with empirically established laws L_A, L_B, and L_C. T_1^* successfully predicts all that T has successfully predicted; T_1^* successfully predicts phen-

late different laws for different phenomena in a "patchwork quilt" fashion are (quite properly) never taken seriously for a moment despite being empirically more successful. This persistent acceptance of unified theories in physics even though endlessly many empirically more successful, patchwork quilt rivals can readily be formulated means that physics makes a persistent assumption about the universe: it is such that all seriously disunified theories are false. The universe is such that some kind of underlying unified pattern of physical law runs through all phenomena.

If physicists only ever accepted theories that postulate atoms even though empirically more successful rival theories are available that postulate other entities such as fields, it would surely be quite clear: physicists implicitly assume that the universe is such that all theories that postulate entities other than atoms are false. Just the same holds in connection with unified theories. That physicists only ever accept unified theories even though endlessly many empirically more successful, disunified rival theories are available means that physics implicitly assumes that the universe is such that all such disunified theories are false.

In accepting the unified theories that it does accept — Newtonian theory, classical electrodynamics, and the rest — physics thereby adopts a big, highly problematic metaphysical hypothesis, H, about the nature of the universe: it is such that all rival, grossly disunified, "patchwork quilt" but empirically more successful theories are false.

H, though a metaphysical hypothesis, is nevertheless a permanent, even if generally unacknowledged, item of theoretical knowledge. Theories that clash with it, even though empirically more successful than accepted physical theories, are

---

omena A that ostensibly refute T; and T_1^* successfully predicts phenomena B and C that T fails to predict. On empirical grounds alone, T_1^* is clearly more successful and better corroborated than T. And all this can be repeated as far as all the other disunified rivals of T are concerned, to generate infinitely many empirically more successful disunified rivals to T: $T_1^*, T_2^*, \ldots T_\infty^*$.

rejected—or rather, are not even considered for acceptance. Whenever a fundamental physical theory is accepted, endlessly many empirically more successful rivals, easily formulated, are not even considered just because, in effect, they clash with H. Thus H is a permanent item of theoretical knowledge in physics, more securely established in scientific practice indeed than any physical theory. Physical theories tend eventually to be shown to be false, but H persists through theoretical revolutions in physics.[34]

Nevertheless, H is an hypothesis, a pure conjecture. How can we make sense of the idea that science is rational and delivers authentic knowledge if the whole enterprise depends crucially on accepting such an unsupported hypothesis as a secure item of scientific knowledge—an hypothesis that exercises a major influence over what theories are accepted and rejected in physics?

6. Aim-Oriented Empiricism

In order to answer this question, we need to adopt a conception of science that I have called *aim-oriented empiricism* (AOE). Precisely because H is a substantial assertion about the nature of the universe, an assertion that, though purely conjectural in character, nevertheless exercises a major influence over what theories are accepted and rejected, even to the extent of overriding empirical considerations, it needs to be made explicit within physics so that it can be critically assessed, rival hypotheses if possible being developed and assessed, in the hope that H can be improved on. We need a new conception of science which represents the metaphysical hypotheses of physics in the form of a hierarchy of hypotheses, as one goes up the hierarchy hypotheses becoming less and less substantial, and more nearly such that their truth is required for science, or the pursuit of knowledge, to be possible at all. In this way we

---

[34] For expositions of this argument see Maxwell (1974, part 1; 1993a, part 1; 1998, ch. 2; 2000c; 2002; 2004a, ch. 1; 2005; 2011b; 2013a).

create a relatively unproblematic framework of hypotheses, and associated methodological rules, high up in the hierarchy, within which much more substantial and problematic hypotheses, and associated methodological rules, low down in the hierarchy, can be critically assessed and, we may hope, improved in the light of the empirical success they lead to and other considerations: see figure 1.

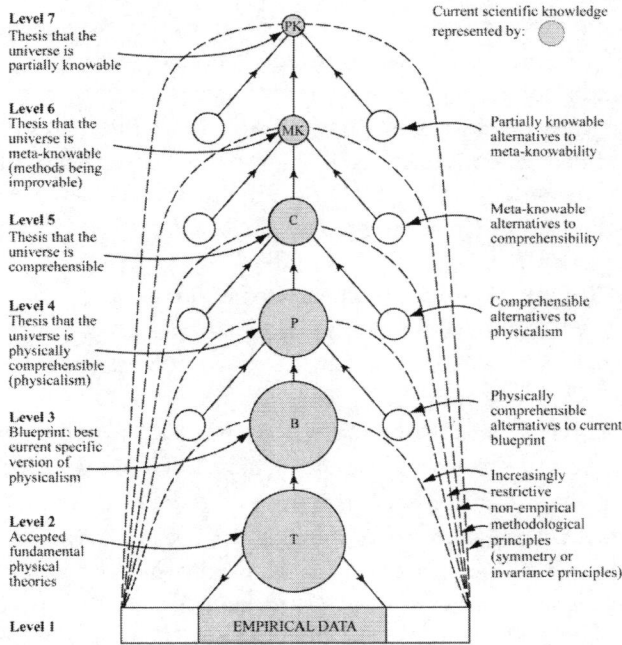

Figure 1: Aim-Oriented Empiricism (AOE)

All this can be reformulated in terms of aims and methods. The aim of science is not truth *per se*, as SE holds. It is rather truth presupposed to be explanatory — or at least knowable. Precisely because this aim of science presupposes a problematic metaphysical hypothesis, the aim (or the hypothesis presupposed by the aim) needs to be represented in the form of a hierarchy of aims (or hypotheses) as indicated in figure 1, so that attempts to improve aims (or hypotheses) may receive the best possible help. As our scientific knowledge and understanding improve, so aims and methods improve as well. There is something like positive feedback between improving scientific knowledge and improving aims and methods—improving knowledge about how to improve knowledge. Science adapts itself to what it finds out about the universe. It is this positive feedback, this interaction between improving scientific knowledge on the one hand, and improving aims and methods (improving assumptions and methods) on the other, that helps explain the explosive growth of modern science. For all this has gone on in scientific practice despite scientists paying lip service to SE. Allegiance to SE has been sufficiently hypocritical to permit aim-oriented empiricism (AOE) to be put into scientific practice, to some extent at least. Allegiance to SE has nevertheless obstructed full implementation of AOE, and has had damaging consequences for science as a result.[35]

There are now three key points to note about AOE.

1. It is not just theoretical physics that has a problematic aim because of problematic hypotheses inherent in the aim. This is true of most—perhaps all—scientific disciplines. Thus most, or perhaps all, scientific disciplines need to be understood in terms of diverse versions of the hierarchical, meta-methodological structure of AOE depicted in figure 1. The aims and methods of science change as we move from one science to another, and as we move within any given science from one time to another. The common factors are (a) something like the hierarchical, interacting structure depicted in figure 1; (b) the common endeavour to improve knowledge and understanding of the universe, and ourselves and other living things as a part of it. AOE provides a general solution to the problem of the nature of the progress-achieving methods of science.[36]

---

[35] For expositions of, and arguments for, AOE see works referred to in note 32 on page 89.
[36] Maxwell (2004a, pp. 39–47).

2. AOE solves fundamental problems in the philosophy of science: in particular, the problem of induction (the problem of the rationality of science); the problem of verisimilitude; and the problem of what it means to say of a physical theory that it is unified.[37]

3. AOE transforms the nature of science, the nature of philosophy of science, and the nature of the relationship between the two. And all this impacts on the nature of the history of science, the sociology of science, and STS. Traditionally, philosophy of science has been conceived of, and practised, as a meta-discipline, studying science in the same way as astronomers study the moon or distant galaxies. This might make sense if science had a fixed aim and fixed methods, as SE holds science does. But AOE asserts that, because the basic aims of science are profoundly problematic, they evolve as scientific knowledge evolves, and change from one science to another. AOE demands that there is a two-way interaction between science itself, on the one hand, and its aims-and-methods, or philosophy, on the other hand. Metaphysics and the philosophy of science become vital ingredients of science itself, concerned to help science make progress. The nature of science, the philosophy of science, and the relationship between the two, all change dramatically.[38]

Exploring probing questions about what the aims of science are, and ought to be, goes much further. For science seeks truth presupposed to be explanatory—explanatory truth as one might say—as a special case of the much more general aim of *valuable truth*—truth that is of intrinsic interest in some way, or of use. A science which increased our knowledge of irredeemably trivial, useless, utterly uninteresting truth would not be said to be making progress. Science both does, and ought to, seek truth that is of use or of value. Merely in order to be accepted for publication, a scientific paper must report a finding that meets

---

[37] Maxwell (1998, chs. 3–6; 2004a, chs. 1, 2, and appendix; 2007a, ch. 14; 2013a).
[38] See works referred to in note 32 on page 89.

some threshold of potential interest. Counting leaves on trees or pebbles on beaches does not, in itself, contribute to scientific knowledge even if the information is new and true.

But the aim of valuable truth is almost more problematic than that of explanatory truth. Of value to whom? And in what way? Is what science seeks to discover always of value to humanity, to those whose needs are the greatest? What of the links that science funding has with the military, corporations of one kind or another, and governments? Do the aims of science always respond to the curiosity and wonder of scientists, or sometimes to their career ambitions and vanity? Given that modern science is expensive, is there not always going to be an inherent conflict between the interests of those who pay for science—the wealthy and powerful—and those whose needs are the greatest—the poor and powerless?

If science is to pursue the problematic aim of valuable truth rationally, and in such a way that justice is done to the best interests of humanity, it is vital that science is pursued within the framework of a generalized version of AOE—humane AOE I have called it—so that three domains of discussion are recognized: (1) evidence; (2) theory; and (3) aims. The third domain of discussion, aims, is as important as the first two. At present it is "repressed"; it goes on in fund-giving committees, and in private between scientists, but not openly in journals and conferences along with (1) and (2). Sustained exploration of the problematic aim of valuable truth needs to attempt to articulate (a) what we conjecture to be scientifically discoverable, and (b) what we conjecture it would be of value to discover, so that we may try to determine the all-important region of overlap between the two. The scientific community may have expertise when it comes to (a), but cannot have any exclusive expertise when it comes to (b). If science is to come to serve the best interests of humanity, it is vital that scientists and non-scientists alike cooperate in engaging in sustained imaginative and critical exploration of what it would be of most value for science to attempt to discover—what ought to be the aims and priorities of scientific and technological research. The institutional/intellectual structure of science needs to be changed to facilitate such aim-exploration. Journals and conferences need to be set up. Science journalism needs to contribute. SE, in misrepresenting the aim of science to be truth *per se*, in effect "represses" the real, problematic aim of valuable truth, and thus damages science by inhibiting the kind of sustained, cooperative exploration of actual and possible valuable aims science does, and might, pursue.[39]

It is important to appreciate that all this comes within the province of philosophy of science which is centrally concerned with problems about the aims and methods of science. Philosophy of science, in order to be done properly, must concern itself with moral, social, value questions about science. It must seek to call into question the less praiseworthy human aspirations science may seek to fulfil — the greed of corporations, the military might of some governments, the self-interests of some scientists. And it must explore neglected avenues of research that might lead to discoveries and technological developments of great potential value to humanity.

It does not stop here. For of course science seeks knowledge of valuable truth so that it may be used by people in life — ideally, so as to enhance and enrich the quality of human life. Science is to be used by people, either culturally, to aid the quest to know, to understand, or practically, as a means to the realization of other goals of value — health, security, travel, communications, entertainment, and so on. Science aims to contribute to the social world. There is a political dimension to the aims of science — once again, profoundly problematic. Everything said above about the value dimensions of the aims of science applies here too to the social, humanitarian, or political dimensions. And this, too, comes within the province of philosophy of science, properly conceived. The orthodox distinction between "internal" factors (purely intellectual) and "external" (social,

---

[39] See my (1976a; 1984; 2001; 2004a; 2007a; 2010a; 2014a).

political, economic, evaluative) is a nonsense. At least, the way this distinction is usually drawn is a nonsense.[40]

7. Damaging Irrationality of Knowledge-Inquiry

We come now to the really substantial step in this exploration of problematic aims. We need to look not just at the aims of natural science and technological science, as we have done so far. In addition, we need to look at the aims of social science too — social science and the humanities, and indeed, the aims of academic inquiry as a whole. The upshot of such an examination of aims is dramatic. We urgently need to bring about a revolution in academic inquiry so that the basic aim becomes wisdom, and not just knowledge.[41]

The official, overall aim of academia, it can generally be agreed, is to help promote human welfare by intellectual and educational means — help people realize what is of value to them in life, help humanity make progress towards as good a world as possible. From the past we have inherited the view that the best way academic inquiry can do this is, in the first instance, to acquire knowledge. First, knowledge has to be acquired; then, secondarily, it can be applied to help solve social problems. Academia organized in this way may be called *knowledge-inquiry*.

Knowledge-inquiry has associated with it a severe censorship system. Only that may enter the intellectual domain of inquiry relevant to the pursuit of knowledge: observational and experimental results, factual claims to knowledge, valid arguments, theories, and so on. Everything else must be ruthlessly excluded: values, feelings and desires, politics, political ideas, policies, cries of distress, problems of living and proposals for

---

[40] See previous note.
[41] The "from knowledge to wisdom" argument I am about to sketch was first expounded in my (1976a). It was spelled out in much greater detail in my (1984); see also my (2007a). See also my (1998; 2001; 2004a; 2010a; 2014a). For summaries that expound different aspects of the argument see Maxwell (1980; 1992; 2000a; 2007b; 2008a; 2011b; 2012a,b).

their solution, philosophies of life—although knowledge about these things can of course be included.

Not everything that goes on in universities today conforms precisely to the edicts of knowledge-inquiry. It is, nevertheless, the dominant view, and exercises a profound influence over what goes on in universities. Knowledge-inquiry is nevertheless profoundly and damagingly irrational in a wholesale, structural way. The irrationality of knowledge-inquiry is so damaging that it is in part responsible for our current incapacity to learn how to tackle effectively our current global problems.

Rationality, as I use the term—and this is the notion that is relevant to the issues we are considering—assumes that there is some probably rather ill-defined set of methods, strategies, or rules which, if put into practice, give us our best chances, other things being equal, of solving our problems or realizing our aims. The rules of reason don't tell us precisely what to do (they tell us what to attempt), and they don't guarantee success. They assume that there is much that we can already do, and they tell us how to marshal these already solved problems in order best to tackle new problems.[42]

There are four elementary rules of reason any problem solving endeavour must implement if it is to be rational, and stand the best chances of meeting with success.

(1) Articulate, and try to improve the articulation of, the basic problem to be solved.

(2) Propose and critically assess possible solutions.

(3) If the basic problem we are trying to solve proves to be especially difficult to solve, specialize. Break the problem up into subordinate problems. Tackle analogous, easier to solve problems in an attempt to work gradually towards the solution to the basic problem.

---

[42] For more about rationality see my (1984, pp. 69–71 and ch. 5; or 2007a, pp. 82–4 and ch. 5).

(4) But if we do specialize in this way, make sure specialized and basic problem solving keep in touch with one another, so that each influences the other.

Any problem solving endeavour that persistently violates just one of these rules will be seriously irrational, and will suffer as a result. Knowledge-inquiry violates *three* of these rules. It is as bad as that.

Knowledge-inquiry puts rule (3) into practice magnificently, especially as exemplified in universities around the world. Endless specialization, disciplines being endlessly subdivided into ever more specialized disciplines, is a striking feature of academia as it exists today. But rules (1), (2), and (4) are all violated.

If we take seriously that academia has as its basic task to help promote human welfare—help people realize what is of value to them in life—then the basic problems academia needs to help solve are *problems of living*, problems of action in the real world, and not, fundamentally, problems of knowledge. It is what we do—or refrain from doing—that enables us to achieve what is of value in life, and not what we know. Even where new knowledge or technology is relevant, as it is in medicine, for example, or agriculture, it is always what this knowledge or technology enables us to do that enables us to achieve what is of value in life, not the knowledge as such (except when knowledge is itself of value).

So, in order to put rules (1) and (2) into practice, academia needs to give absolute intellectual priority to the tasks of (1) articulating our problems of living, including our global problems, and (2) proposing and critically assessing possible solutions—that is, possible *actions, policies, political programmes, strategies, new institutions, new social endeavours, new social arrangements, new ways of living, philosophies of life*. But the censorship system of knowledge-inquiry excludes all this from the intellectual domain of inquiry because it does not constitute contributions to knowledge. Just that which academia most needs to do in order help people, humanity, solve problems of living in increasingly cooperatively rational ways is not done within knowledge-inquiry because it does not contribute to the pursuit of knowledge. And in practice in universities today, thinking about problems of living and policy issues is pushed to the periphery of academia, and does not proceed at the heart of the academic enterprise, as the most fundamental intellectual activity. It is in part because universities today fail to do what most needs to be done to help us make progress towards as good a world as possible that we are in the mess that we are in.

Having violated rules (1) and (2), knowledge-inquiry also violates rule (4). If you fail to engage in thinking about fundamental problems, you cannot interconnect specialized and fundamental problem solving, as rule (4) requires. As a result, specialized research is likely to become unrelated to our most urgent needs which, one may well argue, is what has happened in our universities today.

8. Wisdom-Inquiry: Problem-Solving Version

We need urgently to transform academic inquiry so that all four basic rules of reason are put into practice in a structural way. The outcome is what I have called *wisdom-inquiry*. Wisdom-inquiry is what emerges when knowledge-inquiry is modified just sufficiently to correct its severe rationality defects. At the heart of wisdom-inquiry there are the absolutely fundamental intellectual tasks of (1) articulating and improving the articulation of problems of living, including global problems, and (2) proposing and critically assessing possible solutions — possible actions, policies, political programmes, ways of life, and so on. More specialized problem solving, and in particular scientific and technological research, emerge out of this and feed back into it, in accordance with rules (3) and (4). Thinking about our problems of living and what to do about them influences the aims and priorities of scientific and technological research, and the results of scientific and technological research of course influence thinking about problems of living: see figure 2.

Almost every branch and aspect of academia is modified as we move from knowledge-inquiry to wisdom-inquiry. Within knowledge-inquiry, social inquiry is primarily social science.

The social sciences and humanities have, as their basic task, to improve our knowledge and understanding of social phenomena, the human world. Within wisdom-inquiry, by contrast, the diverse branches of social inquiry have, as their basic task, to articulate problems of living and propose and assess possible solutions. The basic task is to help people, humanity, tackle conflicts and problems of living in the real world in increasingly cooperatively rational ways so that humanity may make progress towards a genuinely good, wise world—or at least as good a world as possible. Social inquiry, so conceived, within wisdom-inquiry, is intellectually more fundamental than natural science.

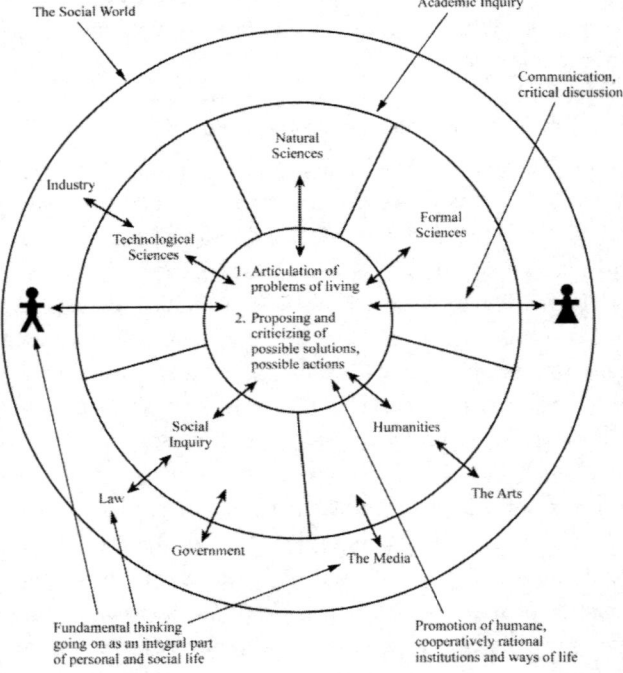

Figure 2: Wisdom-Inquiry Implementing Problem-Solving Rationality

As we move from knowledge-inquiry to wisdom-inquiry the relationship between academia as a whole and the rest of the social world is transformed. Knowledge-inquiry seeks to shield itself from the social world to preserve the objectivity and integrity of the pursuit of knowledge. Wisdom-inquiry, by contrast, seeks to interact with the social world, ideas, experiences, and arguments going in both directions, so that academia may help humanity learn how to tackle our immense global problems more effectively. Wisdom-inquiry might be regarded as a kind of civil service for humanity. What actual civil services are supposed to do in secret for governments, wisdom-inquiry academia does openly for the public.

Knowledge-inquiry has two quite distinct fundamental aims: the intellectual aim of knowledge, and the social or humanitarian aim of helping to promote human welfare. There is a sense in which wisdom-inquiry fuses these together in the one basic aim of seeking and promoting *wisdom*—wisdom being the capacity, and perhaps the active desire, to realize what is of value in life, for oneself and others, wisdom thus including knowledge and technological know-how but much else besides.

9. Wisdom-Inquiry: Aim-Pursuing Version

Granted that the argument of the previous section is correct, and universities today, dominated as they are by knowledge-inquiry, are damagingly irrational in a structural way, an obvious question to ask is: when and how did this come about?

It all goes back to the 18th-century Enlightenment, especially the French Enlightenment. The *philosophes* of the Enlightenment, Voltaire, Diderot, Condorcet, and the rest, had the magnificent idea that it might be possible to learn from scientific progress towards greater knowledge how to make social progress towards an enlightened world. Unfortunately, in developing and implementing this magnificent idea, they blundered. They botched the job. They thought the task was to develop the social sciences alongside the natural sciences. This got developed throughout the 19th century, and got built into universities in the early 20th century with the creation of departments of social science. The outcome is what we have, by and large, today: knowledge-inquiry.

But all this represents a series of dreadful blunders. In order to implement the profound, basic idea of the Enlightenment properly, there are three crucial steps it is essential to get right. The *philosophes* got all three steps wrong.

First, it is essential to get clear about what the progress-achieving methods of science are, what methods, precisely, make scientific progress possible.

Second, these methods need to be correctly generalized so that they become potentially fruitfully applicable to any worthwhile, problematic human endeavour, whatever the aims may be, and not just applicable to the scientific endeavour of improving knowledge.

Third, these correctly generalized progress-achieving methods then need to be got into the social world, into government, industry, agriculture, education, the media, the law, international relations, and so on, so that they may be exploited correctly in the great human endeavour of trying to make social progress towards an enlightened, wise world.

From the 18th century down to today, scientists and philosophers of science have accepted one or other version of standard empiricism (SE) which, as we saw in section 5, very seriously misrepresents the aims and methods of science. In order to get the *first* step right we need to adopt aim-oriented empiricism (AOE).

In order to get the *second* step right, we need to generalize AOE so that it becomes potentially fruitful to any problematic worthwhile human endeavour, and not just science, in this way creating a conception of rationality that helps us improve aims when they are problematic. I have called this aim-pursuing conception of rationality *aim-oriented rationality* (AOR). The vital point to appreciate is that it is not just the aims of science that are problematic; this is true in life as well, in all sorts of personal, social, and institutional contexts. Aims conflict. They have unforeseen, undesirable consequences. They are not as desirable as we suppose, or not as realizable, or both. We may misrepresent our aims. The more "rationally" — that is, effectively — we pursue a bad aim, the worse off we will be. We need to try to improve our aims as we act. Quite generally, whenever we pursue problematic aims, we need to represent them in the form of a hierarchy, along the lines depicted in figure 1, thus giving ourselves the best chances of improving our aims and methods as we act.

In order to get the *third* step right, we need to try to get AOR, arrived at by generalizing AOE, the progress-achieving methods of science, into all our other worthwhile, problematic endeavours besides science — into government, industry, finance, agriculture, education, the media, the law, international relations, and so on. Above all, we need to get AOR into the endeavour to make progress towards the profoundly problematic aim of creating an enlightened world: see figure 3. The *philosophes* made the disastrous mistake of applying a misconceived conception of scientific method, SE, to the task of improving *knowledge* of social phenomena, thus creating social science, when what they ought to have done is apply AOR to *social life itself* so that humanity may make progress towards an enlightened world. According to this second version of wisdom-inquiry (building on the first version), social inquiry is not social *science*, but rather social *methodology* or social *philosophy*. What ought to be the relationship between philosophy of science and science, within the framework of AOE, so too that ought to be the relationship between social inquiry and society. Sociology thus emerges as social *methodology*, and the sociology of science, in particular, emerges as *scientific methodology*, or in other words, *philosophy of science*. At present, philosophy of science and sociology of science are at loggerheads with one another — partly because of social constructivist disagreements. Within this second version of wisdom-inquiry, however, philosophy of science and sociology of science emerge as one and the same discipline, both concerned with what ought to be the intellectual and social aims and methods of science.

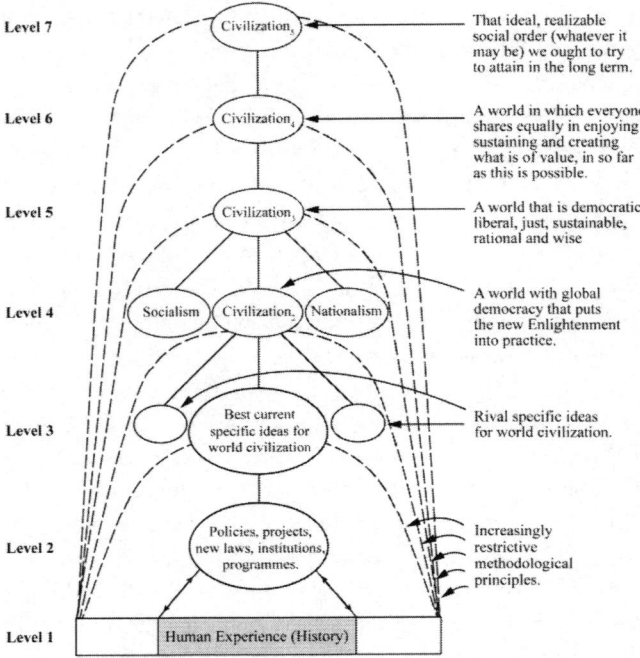

Figure 3: Hierarchical Social Methodology Generalized from Science

10. The Future of Science and Technology Studies

When these arguments for AOE and wisdom-inquiry, just summarized, were spelled out in detail in my book *From Knowledge to Wisdom* in the Orwellian year of 1984, Christopher Longuet-Higgins in a rave review in *Nature* said:

> Maxwell is advocating nothing less than a revolution (based on reason, not on religious or Marxist doctrine) in our intellectual goals and methods of inquiry... There are altogether too many

symptoms of malaise in our science-based society for Nicholas Maxwell's diagnosis to be ignored.[43]

But this is just what has happened. By and large, my diagnosis has been ignored — especially by those who should be most concerned professionally, those engaged in HPS, STS, and philosophy. Instead of bringing to scientists, to fellow academics, and to the public the message that universities, insofar as they put knowledge-inquiry into practice, betray both reason and humanity, these scholars have, rather, devoted themselves to wrangles about social constructivism, and the traditional fair of science studies. The fundamental question *What kind of inquiry can best help humanity learn how to make progress towards as good a world as possible?* continues, scandalously, to be ignored.

I can only hope that this book will provoke some STS folk to take up intellectual cudgels on behalf of reason and humanity.

11. Conclusion

In order to create a better, wiser world, we need to learn how to do it. That in turn requires that our institutions of learning, our schools and universities, are well-designed, rationally designed, and devoted for the task. At present they are not. It is this that is in part responsible for our global problems and our current incapacity to tackle them effectively. We urgently need to bring about a revolution in universities around the world so that they become devoted to seeking and promoting wisdom — helping humanity create a better world. As far as the long term interests of humanity are concerned, there is probably no more important thing that we need to do. Is this academic revolution really needed? What would it imply? What are its advantages and disadvantages? How ought universities to develop? If the revolution is required, what can be done to help bring it about? These are some of the questions STS ought to tackle.

---

[43] Longuet-Higgins (1984).

Chapter Five

Arguing for Wisdom in the University

An Intellectual Autobiography

Introduction

For forty years I have argued that we urgently need to bring about a revolution in academia so that the basic task becomes to seek and promote wisdom. How did I come to argue for such a vast, wildly ambitious intellectual revolution? It goes back to my childhood. From an early age, I desired passionately to understand the physical universe. Then, around adolescence, my passion became to understand the heart and soul of people via the novel. But I never discovered how to tell stories in order to tell the truth. So, having failed to become a physicist, and failed to become a novelist, I studied philosophy at Manchester University and then, in six weeks of inspiration, discovered that the riddle of the universe is the riddle of our desires. Philosophy should be about how to live, and should not just do conceptual analysis. I struggled to reconcile the two worlds of my childhood ambitions, the physical universe and the human world. I decided they could be reconciled with one another if one regarded the two accounts of them, physics and common sense, as myths, and not as literal truths. But then I discovered Karl Popper: truth is too important to be discarded. I revised my ideas: physics seeks to depict truly only an aspect of all that there is; in addition, there is the experiential aspect of things — the world as we experience it. I was immensely impressed with Popper's view that science makes progress not by verification but by ferocious attempted falsification of theories. I was impressed, too, with his generalization of this view to form critical rationalism. Then it dawned on me: Popper's view of science is untenable because it misrepresents the basic aim of science. This is not truth as such; rather it is explanatory truth — truth presupposed to be unified or physically comprehensible. We need, I realized, a new conception of science, called by me *aim-oriented empiricism*, which acknowledges the real, problematic aims of science, and seeks to improve them. Then, treading along a path parallel to Popper's, I realized that aim-oriented empiricism can be generalized to form a new conception of rationality, *aim-oriented rationality*, with implications for all that we do. This led on to a new conception of academic inquiry. From the Enlightenment we have inherited the view that academia, in order to help promote human welfare, must first acquire knowledge. But this is profoundly and damagingly irrational. If academia really does seek to help promote human welfare, then its primary tasks must be to articulate problems of living, and propose and critically assess possible solutions — possible actions, policies, political programmes, philosophies of life. The pursuit of knowledge is secondary. Academia needs to promote cooperatively rational problem solving in the social world, and needs to help humanity improve individual and institutional aims by exploiting aim-oriented rationality, arrived at by generalizing the real progress-achieving methods of science. We might, as a result, get into life some of the progressive success that is such a marked feature of science. Thus began my campaign to promote awareness of the urgent need for a new kind of academic inquiry rationally devoted to helping humanity create a wiser world.

The Key to Wisdom

Nearly forty years ago I discovered a profoundly significant idea — or so I believe. Since then, I have expounded and developed the idea in seven books[1] and countless articles published in academic journals and other books.[2] I have talked about the idea in universities and at conferences all over the UK, in Europe, the USA, Canada, and Taiwan. And yet, alas, despite all this effort, few indeed are those who have even heard of the idea. I have not even managed to communicate the idea to my fellow philosophers.[3]

---

[1] *What's Wrong With Science?*, 1976 & 2009; *From Knowledge to Wisdom*, 1984 & 2007; *The Comprehensibility of the Universe: A New Conception of Science*, 1998, pbk. 2003; *The Human World in the Physical Universe: Consciousness, Free Will and Evolution*, 2001; *Is Science Neurotic?*, 2004; *Cutting God in Half — And Putting the Pieces Together Again*, 2010; *How Universities Can Help Create a Wiser World: The Urgent Need for an Academic Revolution*, 2014.

[2] See, for example, "Science, Reason, Knowledge and Wisdom: A Critique of Specialism", *Inquiry* 23, 1980, pp. 19–81; "What Kind of Inquiry Can Best Help Us Create a Good World?", *Science, Technology and Human Values* 17, 1992, pp. 205–27; "What the Task of Creating Civilization has to Learn from the Success of Modern Science: Towards a New Enlightenment", *Reflections on Higher Education* 4, 1992, pp. 139–57; "Can Humanity Learn to Become Civilized? The Crisis of Science without Civilization", *Journal of Applied Philosophy* 17, 2000, pp. 29–44; "A new conception of science", *Physics World* 13, no. 8, 2000, pp. 17–8; "From Knowledge to Wisdom: The Need for an Academic Revolution", *London Review of Education* 5, 2007, pp. 97–115, republished in Barnett and Maxwell (2008, pp. 1–19); "Do We Need a Scientific Revolution?", *Journal of Biological Physics and Chemistry* 8, no. 3, September 2008, pp. 95–105. All my articles are available online at http://discovery.ucl.ac.uk/view/people/ANMAX22.date.html or http://philpapers.org/profile/17092.

[3] It is not so much the neglect of my work that is the scandal, as the neglect by philosophers of the fundamental problem: *What kind of inquiry can best help us create a good world?* (to quote the title of one of my papers: see Maxwell, 1992). That my work highlights the fundamental significance of this problem — still ignored by academic philosophers today — is almost grounds in itself to think this work merits some attention. But academic philosophers ought, at least, to take *the problem* seriously! That they do not is the real scandal.

What did I discover? Quite simply: the key to wisdom.[4] For over two and a half thousand years, philosophy (which means "love of wisdom") has sought in vain to discover how humanity might learn to become wise — how we might learn to create an enlightened world. For the ancient Greek philosophers, Socrates, Plato, and the rest, discovering how to become wise was the fundamental task for philosophy. In the modern period, this central, ancient quest has been laid somewhat to rest, not because it is no longer thought important, but rather because the quest is seen as unattainable. The record of savagery and horror of the last century is so extreme and terrible that the search for wisdom, more important than ever, has come to seem hopeless, a quixotic fantasy. Nevertheless, it is this ancient, fundamental problem, lying at the heart of philosophy, at the heart, indeed, of all of thought, morality, politics, and life, that I have solved. Or so I believe.

When I say I have discovered the key to wisdom, I should say, more precisely, that I have discovered the *methodological* key to wisdom. Or perhaps, more modestly, I should say that I have discovered that *science* contains, locked up in its astounding success in acquiring knowledge and understanding of the universe, the methodological key to wisdom. I have discovered a recipe for creating a kind of organized inquiry rationally designed and devoted to helping humanity learn wisdom, learn to create a more enlightened world.

What we have is a long tradition of inquiry — extraordinarily successful in its own terms — devoted to acquiring knowledge and technological know-how. It is this that has created the modern world, or at least made it possible. But scientific knowledge and technological know-how are ambiguous blessings, as more and more people, these days, are beginning to recognize. They do not guarantee happiness. Scientific knowledge and technological know-how enormously increase our power to *act*. In endless ways, this vast increase in our power to act has been

---

[4] But see next footnote on page 115.

used for the public good—in health, agriculture, transport, communications, and countless other ways. But equally, this enhanced power to act can be used to cause human harm, whether unintentionally, as in environmental damage (at least initially), or intentionally, as in war. It is hardly too much to say that all our current global problems have come about because of science and technology. The appalling destructiveness of modern warfare and terrorism, vast inequalities in wealth and standards of living between first and third worlds, rapid population growth, environmental damage—destruction of tropical rain forests, rapid extinction of species, global warming, pollution of sea, earth, and air, depletion of finite natural resources— all only exist today because of modern science and technology. Science and technology lead to modern industry and agriculture, to modern medicine and hygiene, and thus in turn to population growth, to modern armaments, conventional, chemical, biological, and nuclear, to destruction of natural habitats, extinction of species, pollution, and to immense inequalities of wealth around the globe.

Science without wisdom, we might say, is a menace. It is the crisis behind all the others. When we lacked our modern, terrifying powers to act, before the advent of science, lack of wisdom did not matter too much: we were bereft of the power to inflict too much damage on ourselves and the planet. Now that we have modern science, and the unprecedented powers to act that it has bequeathed to us, wisdom has become not a private luxury but a public necessity. If we do not rapidly learn to become wiser, we are doomed to repeat in the 21st century all the disasters and horrors of the 20th: the horrifyingly destructive wars, the dislocation and death of millions, the degradation of the world we live in. Only this time round it may all be much worse, as the population goes up, the planet becomes ever more crowded, oil and other resources vital to our way of life run out, weapons of mass destruction become more and more widely available for use, and deserts and desolation spread.

The ancient quest for wisdom has become a matter of desperate urgency. It is hardly too much to say that the future of the world is at stake. But how can such a quest possibly meet with success? Wisdom, surely, is not something that we can learn and teach, as a part of our normal education, in schools and universities?

This is my great discovery! Wisdom *can* be learnt and taught in schools and universities. It *must* be so learnt and taught. Wisdom is indeed the proper fundamental objective for the whole of the academic enterprise: to help humanity learn how to nurture and create a wiser world.

But how do we go about creating a kind of education, research, and scholarship that really will help us learn wisdom? Would not any such attempt destroy what is of value in what we have at present, and just produce hot air, hypocrisy, vanity, and nonsense? Or worse, dogma and religious fundamentalism? What, in any case, *is* wisdom? Is not all this just an abstract philosophical fantasy?

The answer, as I have already said, lies locked away in what may seem a highly improbably place: science! This will seem especially improbable to many of those most aware of environmental issues, and most suspicious of the role of modern science and technology in modern life. How can *science* contain the methodological key to wisdom when it is precisely this science that is behind so many of our current troubles? But a crucial point must be noted. Modern scientific and technological research has met with absolutely astonishing, unprecedented success, as long as this success is interpreted narrowly, in terms of the production of expert knowledge and technological know-how. Doubts may be expressed about whether humanity as a whole has made progress towards well-being or happiness during the last century or so. But there can be no serious doubt whatsoever that science has made staggering intellectual progress in increasing expert knowledge and know-how, during such a period. It is this astonishing intellectual progress that makes science such a powerful but double-edged tool, for good and for bad.

At once the question arises: Can we learn from the intellectual progress of science how to achieve progress in other fields of human endeavour? Is scientific progress exportable, as it were, to other areas of life? More precisely, can the progress-achieving methods of science be generalized so that they become fruitful for other worthwhile, problematic human endeavours, in particular the supremely worthwhile, supremely problematic endeavour of creating a good and wise world?

My great idea — that this can indeed be done — is not entirely new (as I was to learn after making my discovery). It goes back to the 18th-century Enlightenment. This was indeed the key idea of the Enlightenment, especially the French Enlightenment: to learn from scientific progress how to achieve social progress towards an enlightened world. And the *philosophes* of the Enlightenment, men such as Voltaire, Diderot, and Condorcet, did what they could to put this magnificent, profound idea into practice in their lives. They fought dictatorial power, superstition, and injustice with weapons no more lethal than those of argument and wit. They gave their support to the virtues of tolerance, openness to doubt, readiness to learn from criticism and from experience. Courageously and energetically they laboured to promote reason and enlightenment in personal and social life.

Unfortunately, in developing the Enlightenment idea intellectually, the *philosophes* blundered. They botched the job. They developed the Enlightenment idea in a profoundly defective form, and it is this immensely influential, defective version of the idea, inherited from the 18th century, which may be called the "traditional" Enlightenment, that is built into early 21st-century institutions of inquiry. Our current traditions and institutions of learning, when judged from the standpoint of helping us learn how to become more enlightened, are defective and irrational in a wholesale and structural way, and it is this which, in the long term, sabotages our efforts to create a more civilized world, and prevents us from avoiding the kind of horrors we have been exposed to during the last century.

The task before us is thus *not* that of creating a kind of inquiry devoted to improving wisdom out of the blue, as it were, with nothing to guide us except two and a half thousand years of failed philosophical discussion. Rather, the task is the much more straightforward, practical, and well-defined one of *correcting the structural blunders built into academic inquiry inherited from the Enlightenment*. We already have a kind of academic inquiry designed to help us learn wisdom. The problem is that the design is lousy. It is, as I have said, a botched job. It is like a piece of engineering that kills people because of faulty design—a bridge that collapses, or an aeroplane that falls out of the sky. A quite specific task lies before us: to diagnose the blunders we have inherited from the Enlightenment, and put them right.[5]

So here, briefly, is the diagnosis. The *philosophes* of the 18th century assumed, understandably enough, that the proper way to implement the Enlightenment programme was to develop social science alongside natural science. Francis Bacon had already stressed the importance of improving knowledge of the natural world in order to achieve social progress. The *philosophes* generalized this, holding that it is just as important to improve knowledge of the social world. Thus the *philosophes* set about creating the social sciences: history, anthropology, political economy, psychology, sociology.

This had an immense impact. Throughout the 19th century the diverse social sciences were developed, often by non-

---

[5] Note that the outrageous claim with which I began—my claim to have discovered "the key to wisdom"—has become, step by step, very much more modest, even if still a claim concerning a matter of very great importance. I now claim only to have discovered how to correct mistakes made by Enlightenment *philosophes* when they developed their magnificent idea that we need to learn from scientific progress how to achieve social progress towards a wise, enlightened world. Why then did I make the outrageously immodest claim at the beginning of this chapter? It is a desperate attempt to catch the reader's attention. The present chapter represents one more attempt of mine to alert philosophers to the damaging philosophical blunder inherent in the intellectual/ institutional structure of academia as it exists at present, the vital need, for the future of humanity, to bring about an academic revolution so that the basic task becomes to devote reason to helping humanity make progress towards a wise world.

academics, in accordance with this Enlightenment idea. Gradually, universities took notice of these developments until, by the mid-20th century, all the diverse branches of the social sciences, as conceived of by the Enlightenment, were built into the institutional structure of universities as recognized academic disciplines.

The outcome is what we have today, *knowledge-inquiry* as we may call it, a kind of inquiry devoted in the first instance to the pursuit of knowledge.

But, from the standpoint of creating a kind of inquiry designed to help humanity learn how to become enlightened and civilized, which was the original idea, all this amounts to a series of monumental blunders.

In order to implement properly the basic Enlightenment idea of learning from scientific progress how to achieve social progress towards a civilized world, it is essential to get the following three things right.

1. The progress-achieving methods of science need to be correctly identified.

2. These methods need to be correctly generalized so that they become fruitfully applicable to any worthwhile, problematic human endeavour, whatever the aims may be, and not just applicable to the one endeavour of acquiring knowledge.

3. The correctly generalized progress-achieving methods then need to be exploited correctly in the great human endeavour of trying to make social progress towards an enlightened, civilized world.

Unfortunately, the *philosophes* of the Enlightenment got all three points wrong. They failed to capture correctly the progress-achieving methods of natural science; they failed to generalize these methods properly; and, most disastrously of all, they failed to apply them properly so that humanity might learn how to become civilized by rational means. Instead of seeking to apply the progress-achieving methods of science, after having been appropriately generalized, to the task of creating a better world, the *philosophes* applied scientific method to the task of creating social *science*. Instead of trying to make *social* progress towards an enlightened world, they set about making *scientific* progress in knowledge of social phenomena. That the *philosophes* made these blunders in the 18th century is forgivable; what is unforgivable is that these blunders still remain unrecognized and uncorrected today, over two centuries later. Instead of correcting them, we have allowed our institutions of learning to be shaped by them as they have developed throughout the 19th and 20th centuries, so that now the blunders are an all-pervasive feature of our world.

The Enlightenment, and what it led to, has long been criticized, by the Romantic movement, by what Isaiah Berlin has called 'the counter-Enlightenment', and more recently by the Frankfurt school, by postmodernists, and others. But these standard objections are, from my point of view, entirely missing the point. In particular, my idea is the very opposite of all those anti-rationalist, romantic, and postmodernist views which object to the way the Enlightenment gives far too great an importance to natural science and to scientific rationality. My discovery is that what is wrong with the traditional Enlightenment, and the kind of academic inquiry we now possess derived from it — *knowledge-inquiry* — is not too much 'scientific rationality' but, on the contrary, not enough. It is the glaring, wholesale *irrationality* of contemporary academic inquiry, when judged from the standpoint of helping humanity learn how to become more civilized, that is the problem.

But, the cry will go up, wisdom has nothing to do with reason. And reason has nothing to do with wisdom. On the contrary! It is just such an item of conventional 'wisdom' that my great idea turns on its head. Once both reason and wisdom have been rightly understood, and the irrationality of academic inquiry as it exists at present has been appreciated, it becomes obvious that it is precisely *reason* that we need to put into practice in our personal, social, institutional, and global lives if our lives, at all these levels, are to become imbued with a bit more wisdom. We need, in short, a new, more rigorous kind of inquiry which has, as its basic task, to seek and promote *wisdom*. We may call this new kind of inquiry *wisdom-inquiry*.

But what is wisdom? This is how I define it in *From Knowledge to Wisdom*, a book published some years ago now, in 1984, in which I set out my 'great idea' in some detail:

> [Wisdom is] the desire, the active endeavour, and the capacity to discover and achieve what is desirable and of value in life, both for oneself and for others. Wisdom includes knowledge and understanding but goes beyond them in also including: the desire and active striving for what is of value, the ability to see what is of value, actually and potentially, in the circumstances of life, the ability to experience value, the capacity to use and develop knowledge, technology and understanding as needed for the realization of value. Wisdom, like knowledge, can be conceived of, not only in personal terms, but also in institutional or social terms. We can thus interpret [wisdom-inquiry] as asserting: the basic task of rational inquiry is to help us develop wiser ways of living, wiser institutions, customs and social relations, a wiser world.[6]

What, then, are the three blunders of the Enlightenment, still built into the intellectual/institutional structure of academia?

First, the *philosophes* failed to capture correctly the progress-achieving methods of natural science. From d'Alembert in the 18th century to Karl Popper in the 20th, the widely held view, amongst both scientists and philosophers, has been (and continues to be) that science proceeds by assessing theories impartially in the light of evidence, *no permanent assumption being accepted by science about the universe independently of evidence*. Preference may be given to simple, unified, or explanatory theories, but not in such a way that nature herself is, in effect, assumed to be simple, unified, or comprehensible.

This orthodox view, which I call *standard empiricism* is, however, untenable. If taken literally, it would instantly bring science to a standstill. For, given any accepted fundamental theory of physics, T, Newtonian theory say, or quantum theory,

---

[6] See Maxwell (1984, p. 66; 2007a, p. 79).

endlessly many empirically more successful rivals can be concocted which agree with T about observed phenomena but disagree arbitrarily about some unobserved phenomena, and successfully predict phenomena, in an *ad hoc* way, that T makes false predictions about, or no predictions. Physics would be drowned in an ocean of such empirically more successful rival theories.

In practice, these rivals are excluded because they are disastrously disunified. *Two* considerations govern acceptance of theories in physics: empirical success and unity. In demanding unity, we demand of a fundamental physical theory that it ascribes *the same* dynamic laws to the phenomena to which the theory applies.[7] But in persistently accepting unified theories, to the extent of rejecting disunified rivals that are just as, or even more, empirically successful, physics makes a big persistent assumption about the universe. The universe is such that all disunified theories are false. It has some kind of unified dynamic structure. It is physically comprehensible in the sense that explanations for phenomena exist to be discovered.

But this untestable (and thus metaphysical) assumption that the universe is physically comprehensible is profoundly problematic. Science is obliged to assume, but does not know, that the universe is comprehensible. Much less does it know that the universe is comprehensible in this or that way. A glance at the history of physics reveals that ideas have changed dramatically over time. In the 17th century there was the idea that the universe consists of corpuscles, minute billiard balls, which interact only by contact. This gave way to the idea that the universe consists of point-particles surrounded by rigid, spherically symmetrical fields of force, which in turn gave way to the idea that there is one unified self-interacting field, varying smoothly throughout space and time. Nowadays we have the idea that

---

[7] For a more detailed account of this conception of the unity of theory, see Maxwell (1998, especially ch. 4; 2004a, appendix, section 2; 2007a, ch. 14, section 2).

everything is made up of minute quantum strings embedded in ten or eleven dimensions of space-time. Some kind of assumption along these lines must be made but, given the historical record, and given that any such assumption concerns the ultimate nature of the universe, that of which we are most ignorant, it is only reasonable to conclude that it is almost bound to be false.

The way to overcome this fundamental dilemma inherent in the scientific enterprise is to construe physics as making a hierarchy of metaphysical assumptions concerning the comprehensibility and knowability of the universe, these assumptions asserting less and less as one goes up the hierarchy, and thus becoming more and more likely to be true, and more nearly such that their truth is required for science, or the pursuit of knowledge, to be possible at all. In this way a framework of relatively insubstantial, unproblematic, fixed assumptions and associated methods is created within which much more substantial and problematic assumptions and associated methods can be changed, and indeed improved, as scientific knowledge improves. Put another way, a framework of relatively unspecific, unproblematic, fixed *aims* and methods is created within which much more specific and problematic aims and methods evolve as scientific knowledge evolves. There is positive feedback between improving knowledge, and improving aims-and-methods, improving knowledge-about-how-to-improve-knowledge. This is the nub of scientific rationality, the methodological key to the unprecedented success of science. Science adapts its nature to what it discovers about the nature of the universe. Philosophy of science (the study of the aims and methods of science) becomes an integral, vital part of science itself. And science becomes much more like natural philosophy in the time of Newton, a synthesis of science, methodology, epistemology, metaphysics, and philosophy.

This hierarchical conception of physics, which I call *aim-oriented empiricism*, can readily be generalized to take into account problematic assumptions associated with the aims of science having to with *values*, and the *social uses* or *applications* of science. It can be generalized so as to apply to the different branches of natural science. Different sciences have different specific aims, and so different specific methods, although throughout natural science there is the common meta-methodology of aim-oriented empiricism.

So much for the first blunder of the traditional Enlightenment, and how to put it right.[8]

Second, having failed to identify the methods of science correctly, the *philosophes* naturally failed to generalize these methods properly. They failed to appreciate that the idea of representing the problematic aims (and associated methods) of science in the form of a hierarchy can be generalized and applied fruitfully to other worthwhile enterprises besides science. Many other enterprises have problematic aims—problematic because aims conflict, and because what we seek may be unrealizable, undesirable, or *both*. Such enterprises, with problematic aims, would benefit from employing a hierarchical methodology, generalized from that of science, thus making it possible to improve aims and methods as the enterprise proceeds. There is the hope that, as a result of exploiting in life methods generalized from those employed with such success in science, some of the astonishing success of science might be

---

[8] For further details see my *The Comprehensibility of the Universe*, 1998; *Is Science Neurotic?*, 2004; *From Knowledge to Wisdom*, especially chs. 5, 9, and 2nd ed., ch. 14; "Popper, Kuhn, Lakatos and Aim-Oriented Empiricism", *Philosophia* 32, nos. 1–4, 2005, pp. 181–239; "The Metaphysics of Science: An Account of Modern Science in Terms of Principles, Laws and Theories" (review of book by Craig Dilworth), *International Studies in the Philosophy of Science* 23, no. 2, 2009, pp. 228–32; *Cutting God in Half—And Putting the Pieces Together Again: A New Approach to Philosophy*, 2010 (available free online at http://discovery.ucl.ac.uk/view/people/ANMAX22.date.html); "Has Science Established that the Cosmos is Physically Comprehensible?", in Travena, A. and Soen, B., eds., *Recent Advances in Cosmology*, 2013; "Unification and Revolution: A Paradigm for Paradigms", *Journal for General Philosophy of Science*, 2014, http://philpapers.org/rec/MAXUAR. See also note 34 on page 165–6.

exported into other worthwhile human endeavours, with problematic aims quite different from those of science.

Third, and most disastrously of all, the *philosophes* failed completely to try to apply such generalized, hierarchical progress-achieving methods to the immense, and profoundly problematic enterprise of making social progress towards an enlightened, wise world. The aim of such an enterprise is notoriously problematic. For all sorts of reasons, what constitutes a good world, an enlightened, wise, or civilized world, attainable and genuinely desirable, must be inherently and permanently problematic. Here, above all, it is essential to employ the generalized version of the hierarchical, progress-achieving methods of science, designed specifically to facilitate progress when basic aims are problematic. It is just this that the *philosophes* failed to do. Instead of applying the hierarchical methodology to *social life*, the *philosophes* sought to apply a seriously defective conception of scientific method to *social science*, to the task of making progress towards not a *better world* but to better *knowledge* of social phenomena. And this ancient blunder, developed throughout the 19th century by J.S. Mill, Karl Marx, and many others, and built into academia in the early 20th century with the creation of the diverse branches of the social sciences in universities all over the world, is still built into the institutional and intellectual structure of academia today, inherent in the current character of social science.

Properly implemented, in short, the Enlightenment idea of learning from scientific progress how to achieve social progress towards an enlightened world would involve developing social inquiry, not primarily as social *science*, but rather as social *methodology*, or social *philosophy*. A basic task would be to get into personal and social life, and into other institutions besides that of science – into government, industry, agriculture, commerce, the media, law, education, international relations – hierarchical, progress-achieving methods (designed to improve problematic aims) arrived at by generalizing the methods of science. A basic task for academic inquiry as a whole would be to help humanity learn how to resolve its conflicts and problems of living in more just, cooperatively rational ways than at present. The fundamental intellectual and humanitarian aim of inquiry would be to help humanity acquire wisdom—wisdom being, as I have already indicated, the capacity to realize (apprehend and create) what is of value in life, for oneself and others.

One outcome of getting into social and institutional life the kind of aim-evolving, hierarchical methodology indicated above, generalized from science, is that it becomes possible for us to develop and assess rival philosophies of life as a part of social life, somewhat as theories are developed and assessed within science. Such a hierarchical methodology provides a framework within which competing views about what our aims and methods in life should be—competing religious, political, and moral views—may be cooperatively assessed and tested against broadly agreed, unspecific aims (high up in the hierarchy of aims) and the experience of personal and social life. There is the possibility of cooperatively and progressively improving *such philosophies of life* (views about what is of value in life and how it is to be achieved) much as *theories* are cooperatively and progressively improved in science.

Wisdom-inquiry, because of its greater rigour, has intellectual standards that are, in important respects, different from those of knowledge-inquiry. Whereas knowledge-inquiry demands that emotions and desires, values, human ideals and aspirations, philosophies of life be excluded from the intellectual domain of inquiry, wisdom-inquiry requires that they be included. In order to discover what is of value in life it is essential that we attend to our feelings and desires. But not everything we desire is desirable, and not everything that feels good is good. Feelings, desires, and values need to be subjected to critical scrutiny. And of course feelings, desires, and values must not be permitted to influence judgments of factual truth and falsity.

Wisdom-inquiry embodies a synthesis of traditional Rationalism and Romanticism. It includes elements from both, and it improves on both. It incorporates Romantic ideals of integrity, having to do with motivational and emotional honesty, honesty about desires and aims; and at the same time it incorporates traditional Rationalist ideals of integrity, having to do with respect for objective fact, knowledge, and valid argument. Traditional Rationalism takes its inspiration from science and method; Romanticism takes its inspiration from art, from imagination, and from passion. Wisdom-inquiry holds art to have a fundamental rational role in inquiry, in revealing what is of value, and unmasking false values; but science, too, is of fundamental importance. What we need, for wisdom, is an interplay of sceptical rationality and emotion, an interplay of mind and heart, so that we may develop mindful hearts and heartfelt minds (as I put it in my first book *What's Wrong With Science?*). It is time we healed the great rift in our culture, so graphically depicted by C.P. Snow.[9]

The revolution we require—intellectual, institutional, and cultural—if it ever comes about, will be comparable in its long term impact to that of the Renaissance, the scientific revolution, or the Enlightenment. The outcome will be traditions and institutions of learning rationally designed to help us realize what is of value in life. There are a few scattered signs that this intellectual revolution, from knowledge to wisdom, is already under way.[10] It will need, however, much wider cooperative support—from scientists, scholars, students, research councils, university administrators, vice chancellors, teachers, the media, and the general public—if it is to become anything more than what it is at present, a fragmentary and often impotent movement of protest and opposition, often at odds with itself, exercising little influence on the main body of academic work. I can hardly imagine any more important work for anyone associated with academia than, in teaching, learning, and research, to help promote this revolution.

---

[9] Snow (1964).
[10] See my (2007a, ch. 12; 2009b; 2014a, ch. 4).

Childhood

It may be thought that my idea that I can publish a few books and articles, give a few lectures, and thereby, single-handedly as it were, transform the entire academic enterprise, amounts to megalomania, if not downright lunacy. Where did such a mad project come from?

It all goes back to my childhood. For as far back as I can remember, I seem to have found the world baffling, mysterious, and frightening. Above all I was terrified by the black inevitability of death. From the age of four, I was haunted by problems of war, theology, cosmology, physics, consciousness, epistemology, and the meaning of life.

One night, when I was three years old, during the early stages of the Second World War, the German Luftwaffe dropped bombs in a field not so very far from our house. Later that night I paced to and fro in my parents' bedroom, my hands deep in my dressing gown pockets, my head bowed in thought. Finally, I stopped, turned to my parents, and asked: "Mummy, why do they have wars?" Today, I am proud of my three-year-old self for asking that good question.

Around the same time, I entered into a fierce theological debate with the boy next door. He was trying to convince me of the existence of God. "If God doesn't exist", he argued, "who do you think made the earth? Who made the trees? Who made the stars?" I listened to this litany of questions in silence for a while, and then asked in turn: "And who made God?" The boy next door went away without giving me an answer.

A little later, when I was four, I got interested in natural philosophy and cosmology. I invented a theory as to why the sky is blue. According to this theory, the sky is blue because air is very, very slightly blue. When you look at things close to, the blueness of the air is too slight to be noticed, but when you look at the sky, you see through so much air that the blueness is easy to see. I can remember trying to convince my father of this explanation for the blueness of the sky — and I remember my exasperation when, for some extraordinary reason, he remained unconvinced.

I also remember lying awake in bed one summer evening at this time, puzzling about how space can come to an end. It occurred to me that far away in the sky there must be a vast wall that marks the outer boundary of everything. For a while, this seemed to me to be a satisfactory enough solution to the problem. And then I had the awful thought: but what is behind the wall? Something must be behind the wall!

About a year later, when I was about five, I made the extraordinary discovery of self-consciousness. I had had a row with my mother. She wanted me and my sister to go for a walk. I protested. "It will rain", I declared, pointing to some dark clouds. Off my mother and sister went, leaving me behind. Feeling somewhat resentful and self-righteous, alone in the house, it began to dawn on me that I had something infinitely precious and mysterious that no one else had: my own awareness of myself, my inner secret thoughts and feelings.

A year later, by the age of six, my passion for natural philosophy was well aroused. One day, I asked my father how it was possible to make tubes as small as those in the filaments of electric light bulbs, so that electricity could flow through. My father explained that the filaments, like all electric wires, are solid metal. At first I was furiously indignant: how could electricity possibly flow through solid metal? But when it became clear that my father really did know what he was talking about, I fell silent, stupefied by this mystery of electricity flowing through solid metal. Electricity took on for me a quality that was both fascinating and nightmarish. I knew it was dangerous, and could kill. I had been told that in an unused, upper story of our house, firmly out of bounds, there were uninsulated "live" wires. I imagined "live" wires lashing out, dealing out their terrible sting of electric death. On one occasion a girl visiting for the day, much older than me (she was eight or nine years old) persuaded me, against my better judgment, to put my finger into the empty socket of a table lamp. She assured me that it was perfectly safe and that I would not feel a thing. In fact I received a shock—fortunately only through the tip of my finger. Here, then, was the violent rushing pain and mystery of electricity, experienced at first hand. From a cautious distance, I would contemplate the transformer in our garden, hidden behind some bushes, softly humming to itself, quietly containing its secret, deadly power. At about this time, I began to take torches and batteries to pieces to try to discover how they worked.

One day while in the garden, I made what seemed to me to be a wonderful discovery. I discovered a reason for believing in the existence of atoms. If atoms did not exist—I felt rather than thought—and matter remained exactly the same, however minutely it might be subdivided, then there could exist nothing to fix the size of things. Things could be any size. But things are not any size: somehow, people, animals, plants do know roughly what size to be. Therefore atoms of some kind or other, of a definite size, must exist, to fix the size of everything else.

I found this argument entirely convincing, although not for one moment did I suppose it would convince anyone else. Indeed, the dramatic and extraordinary discovery that I felt I had made was, at the time, wholly private, uncommunicable, beyond words, my own personal wordless recognition of the force of the argument that I have here spelled out in words, a feeling rather than a thought. I did not imagine at the time that an insight so emotional and personal could be put into words, and thus be rendered open to public understanding and scrutiny.

Also, at about this time (around the age of five or six), I discovered for myself the problem of perception. I was sitting on the sofa in the living room, and I began to think about what was going on as I looked about me at objects in the room. I thought about the light which was reflected from tables, chairs, the walls of the room and which then entered my eyes to cause me to have the experience of seeing. What I was really seeing, it seemed, was the light entering my eyes, not the furniture around me. Here was the sofa, the carpet, the table, wholly visible and obvious before me. And yet, it seemed, I could not possibly be seeing these things. I could only really see what happens when light enters my eyes. This room I was seeing must somehow be inside my head—and yet it could not possibly be inside my head! The more I thought about it the more horrible the problem became. Mentally, if not physically, I was staggering about the room, clutching my head, tearing out my hair, bewildered beyond belief.

In recounting these childhood discoveries (in the main discoveries of problems rather than of solutions to problems), I am perhaps in part just boasting, in a rather foolish and shameful way. Certainly, I am today absurdly proud of these childhood discoveries of mine: I tremble to think of how I may subsequently have squandered the early passionate intellectual curiosity and independence these discoveries reveal. But in another way, I am not boasting at all. All of us, I believe, are extraordinarily active and creative intellectually when we are very young. Bryan Magee gives a vivid, dramatic account of similar philosophical discoveries that he made when young.[11] Somehow, in the first few years of life, we acquire an identity, a consciousness of self; we discover, or create, a whole view of the world, a cosmology; and we learn to understand speech, and to speak ourselves. And we achieve all this without any formal education whatsoever. Compared with these mighty intellectual achievements of our childhood, the heights of adult artistic and scientific achievement all but fade into insignificance. It is reasonable to suppose that there is a biological, a neurological, basis for our extraordinary capacity to learn when we are very young. It probably has to do with the fact that our brains are still growing during the first few years of our life. It is striking that there are things that can only be learnt during this time. If we have not had the opportunity to learn to speak by the age of twelve, we will never really learn to speak. Lightning calculators all begin to acquire their extraordinary arithmetical skills when very young. Some things it seems become too difficult for

---

[11] See Magee (1997, ch. 1: Scenes from Childhood). The impulse to write up my childhood "discoveries" owes nothing to Magee's account; it was written long before publication of Magee's book, sometime before 1987.

us to learn as we grow older. In our early childhood we are forced, by our situation, to be creative philosophers and metaphysicians, preoccupied by fundamental issues. One only has to think of the endless questioning of young children to appreciate something of their insatiable hunger to know, to understand.

The tragedy is that formal education fails so dismally to recognize, and to help nourish, this frenzy of childish curiosity. At school we are expected to learn up items of human knowledge—solutions to other people's problems. It is rather rare to be told about, or to be asked to consider—let alone to be encouraged to wrestle with—the problems which gave rise to these solutions: and yet only this can enable us to make rational sense of the solutions themselves. It is even rarer to be asked to articulate our own problems, and our ideas as to how they might be solved. Worst of all, much education, unintentionally, makes us ashamed of our own intellectual integrity and creativity. At school a premium is placed on being able to understand quickly, and remember. We thus tend to grow ashamed of what we take to be our "stupidity"—our inability to understand, our puzzlement, our incomprehension. And yet it is precisely here, in our inability to understand, our sensitivity to the existence of problems, that our real intelligence and integrity lie. In prompting us to disown our inner stupidity, our lingering sense of bafflement, education encourages us to disown the precious core of our mind.

The result is that we come to devalue and forget our childhood discoveries. We do not appreciate even that the discovery of a problem can be a great intellectual achievement. We do not learn how to translate feelings of bafflement into articulated questions, into public words: and so memory of the bafflement is lost.[12]

---

[12] I suggested in Chapter One how education can be reformed so that it stimulates and builds on childish curiosity and does not crush it out of existence. Also relevant is my (1980).

The Physical Universe

By the time I was eight, my parents had decided, perhaps with a touch of amusement, that I was to be a "scientist". In those days, at the end of the Second World War, to be a scientist was considered to be a highly desirable and honourable profession to aspire to—for a boy at least, and only if one was clever enough. In our family, however, science carried with it no special status or prestige: that went rather to literature, to the arts, to the creative and fulfilling life. But as far as I was concerned, it was not any kind of profession that was on my mind at all. My ambition, quite simply, was to solve the ultimate riddle, understand the ultimate nature of the universe, the nature of existence. To live and to die, and not know what kind of world this is, what it all means, seemed to me then, at the age of about eight, to be a fate too terrible to think of. Yet this, unfortunately, was the fate of everyone who had lived up until now. No one, I was convinced, had ever had the faintest idea of the true nature of the cosmos, the true inner meaning of it all. Most people were not even aware of the disaster of their ignorance. They lived and died unaware of the tragic triviality and irrelevance of their lives. Life could only acquire its real meaning if one could clearly see and know. I had no choice: I must know and understand, as a matter of necessity. Where everyone else had failed, I must succeed. And when I discovered the great secret of the inner nature of the universe, I would reveal it to mankind, and be loved forever.

At the age of ten, fired by this mighty and terrible ambition —in effect to become the saviour of mankind—I plunged into the study of nuclear physics! I devoured the contents of *Science News 2* (Peierls & Enogat, 1947), which was devoted entirely to an informal account of nuclear energy and the bomb, with contributions from Peierls, Bethe, Teller, Frisch, and others. I understood enough to be worried about the possibility that an atomic bomb, exploded in the ocean, might turn the earth into another sun. In the great heat and pressure of the sun, I read, atoms of heavy hydrogen combine to form helium, a reaction that causes the sun to shine. If it happens there, I thought, it could happen here as well. And quite apart from that horror, there was the possibility of atomic war to think about. I made anxious calculations about our chances of survival if a bomb was dropped on Truro or Bodmin, living as we did on the north coast of Cornwall. Exploration of the inner secrets of nature brought with it, it seemed, both wonder and terror.

A year later I plunged into the study of relativity and quantum theory. With fascinated incomprehension I read Bertrand Russell's *ABC of Relativity* and Eddington's *The Nature of the Physical Universe*. I learned that as we move more quickly, we shrink, time goes more slowly, and we become more massive. Gravitation is simply the curvature of space-time. Everything is made up of electrons, protons, and neutrons; but these fundamental entities, even though particles, are also, in some utterly mystifying way, wave-like in character as well, waves of probability. As a result of the investigations of science, the solid and prosaic world around us is revealed to be something utterly different, a place of dark miracles and mystery. It was above all my imagination that was appealed to by the utter strangeness of this world disclosed to us by modern physics. And buried within this mystery, this jabberwocky world, lay the solution to the enigma of existence. My intentions were, it seemed, becoming clearer. I would be a theoretical physicist, and discover the solution to the ultimate riddle of existence.

Around this time I read my first book of philosophy: W.A. Sinclair's *An Introduction to Philosophy*. This I read with interest, struck especially by the brief account of Hume's argument concerning the impossibility of knowing for sure that the sun will rise tomorrow, however many times it may have risen regularly in the past and however firmly our most successful scientific theories might predict the occurrence. I realized, embarrassed, that I had somehow supposed that the authoritative body of scientists must be in a position to know such an obvious fact about the world in a way which placed it beyond all doubt — the universe, as it were, not daring to disobey the weighty judgment of the adult world. How absurd! Of course it could not be like that. It must of course always be possible for the universe to surprise people, however convinced they might be that this could not occur, however stiff and dignified they might be with certainty. I felt ashamed of my gullibility, and also interested that such an elementary argument could have such scope, such power to change the way one viewed things.

On the whole, however, I was not very impressed with philosophy. At the time, and for some years afterwards, it struck me as a game rather than as something serious. Not for one moment did I suppose that the solution to the mystery of existence, which I sought, could lie hidden in something as feeble as philosophy.

None of this, by the way, should be taken to mean that I was horribly precocious. Not at all. In those far off days in England, 11-year-olds had to take an exam which decided whether they would be able to go on to grammar school or not. Failure to pass this exam more or less condemned you to leaving school without qualifications (unless your parents could pay for your education). I failed this crucial exam, not once, but twice!

The Human World

Then, with the arrival of the traumas, ecstasies, and disasters of adolescence, I began to feel it was much more important to understand the hearts and souls of people, and the way to do that was by means of the novel. Instead of reading Jeans, Eddington, and Fred Hoyle, I plunged into the worlds of Dostoevsky, Jane Austen, Henry Fielding, Chekhov, Stendhal, D.H. Lawrence, Kafka, Virginia Woolf, Scott Fitzgerald, Thomas Mann, Tolstoy, Balzac, George Orwell, Turgenev, Conrad, Thomas Hardy, Ibsen, James Joyce, Mauriac, Bernard Shaw, George Eliot, Emile Bronte. I read *The Brothers Karamazov* in two days, emerging briefly, dazed and battered, from that turbulent and tortured world into thin reality for a bite to eat at lunch and supper. One afternoon I took a slim book by an unknown author up to my bedroom to read at a sitting; I was so astonished by its contents that I returned it hurriedly to the bookshelf, making sure no one noticed, as if the book were an obscene publication. It was Kafka's *Metamorphosis*. I hunted for books which would open up new worlds, so intensely imagined and so truthful that they would seem more vivid, more dense and real than the real world itself. What I wanted was not just the accurate depiction of this world, but the creation of a new, strange world experienced as reality. I marvelled at the early pages of Virginia Woolf's *The Waves*, Kafka's *The Trial*, Emily Brontë's *Wuthering Heights*, Dostoevsky's *The Idiot* and *The Brothers Karamazov*. Science fiction was for me only a cheap thrill. I read H.G. Wells' *The Time Machine*, and Stevenson's *Dr. Jekyll and Mr. Hyde*, gripped and fascinated: but for me neither book even began to engage in the proper task of the novel. What I sought was an exploration of the realities of human experience and emotion that was so truthful, so searching and profound, that we are led when we encounter it into a new vision of reality, a vision before only dimly and fleetingly sensed, so that we encounter it now in an overwhelming way as both strange and familiar, a wide awake dream. It occurs to me now that what I wanted was to be shown a world that was both as real and as mysterious as the universe of modern physics; but the reality and the mystery should lie in the human mind and heart, in our inner lives, and only incidentally, as it were, in any rearranging of the outer cosmic order. It was as if I believed reality must be utterly mysterious—whether the reality of the physical universe, or the reality of the human soul. Something of this I found in Kafka, and in Dostoevsky—and later on in Strindberg, and in the best films of Ingmar Bergman. And of course it is to be found in Shakespeare. And in Beckett.

My great ambition was transformed. By the age of fifteen I had no doubts, I would become a novelist. Clear sighted and unflinching, I would journey into the depths of the human heart and mind, into that cauldron of desire and terror, fantasy and nightmare we often pretend does not exist. I would capture the very essence of what it is to be conscious and alive—the intense inner feeling of ourself which we all know but do not know how to express. I would come up with the true inner meaning and value of our lives, the precious essence of life. And the novels I would write would be revelations of these inner realities—so intense, vivid, and dramatic as to be more real than reality itself.

My parents, however, insisted that, first, I must go to university, to secure my future economically (of no significance to me at all at the age of 17). The educational system, fiercely classificatory in those days, had labelled me "science" and not "humanities". (And in any case I knew doing English at university would ruin any chance I might have of becoming a novelist.) I had read Eddington, who informed me that physics is really mathematics, and for a time, earlier, I had been dazzled by this invisible, esoteric world of mathematics. So off I went to University College London to do mathematics, convinced I could write my novels between and after lectures.

But I was miserable; I didn't know what to write about; and mathematics seemed both hollow and very difficult. It did not seem to be about anything. I passed all my exams but, abruptly, in my second year, my grant was stopped because I had not attended enough lectures.

So I left and did my National Service in Bielefeld, Germany. I became a Sergeant in the Educational Corps. And then I went to Manchester University to do philosophy. I had failed miserably as a physicist, and as a novelist, but I was interested in philosophical problems, so I would do that for three years, and then join the grey shuffle of ordinary, uncreative life (as I then saw it).

But before I plunge into an account of what happened as a result of going to Manchester, there is one other influence from my childhood that I must mention. The household god was Sigmund Freud. My mother had been psychoanalysed. Freud informed her vision of the world. But what I learned from her about Freud I found deeply disturbing. If my unconscious controlled my actions, this meant *I* was not in control. Free will was an illusion. Freud had to be refuted. But there did seem to be something to Freudianism. All too often what was supposed to be going on in human affairs seemed to me to be at odds with what was really going on. I decided the only way to refute Freud was never to deny an interpretation of my actions, motivations, and feelings, however devastating that interpretation might be. As a result, my unconscious would gradually become conscious, and I would regain control of my actions and my life. Perhaps my adoption of this strategy to refute Freud accounts at least in part for the excesses that are to follow.[13]

Manchester University

At Manchester, in the first year, there were just two courses, both introductory: logic and philosophy. While still in Germany, I knew I would be doing something called "symbolic logic" at Manchester, but I had no idea what it was, and there were no books available to tell me. If I was to find out what it was, I was going to have to reinvent it myself. So I got hold of a big yellow army exercise book, and filled it with my efforts to reinvent symbolic logic. For months, I struggled to put Aristotelian logic into symbolic form, but got nowhere.

Then, at Manchester, I was introduced to the propositional calculus, and I was enchanted. It had never occurred to me to develop symbolic logic in such a fashion. The others doing the course were bored, but I was entranced and alive with questions. This episode came to dramatize for me what is so tragically wrong with so much education, at all levels. Our heads are stuffed with solutions to problems. Rarely are we told what the problems are in the first place — something we need to know to be able to assess, for ourselves, how adequate the proposed

---

[13] That my work has been influenced by Freud is obvious from the title alone of one of my books: *Is Science Neurotic?* I have argued, however, that Freud needs to be reinterpreted radically, so that he becomes a *methodologist*. (Psychoanalytic theory itself suffers from rationalistic neurosis.) This reinterpretation transforms the quarrel between Freud and science, highlighted by Popper, Grünbaum, and others. "...It is not Freud who fails to match up to the exacting standards of science; on the contrary, it is *science* that fails to match up to the exacting intellectual standards of Freudianism reinterpreted methodologically. Science suffers from rationalistic neurosis, and needs methodological treatment": *Is Science Neurotic?*, p. 111. See also my (1984, pp. 110–7; 2nd ed., 2007a, pp. 122–9).

solutions are. Rarely indeed are we given the opportunity to struggle ourselves to attempt to solve fundamental intellectual problems—not with the idea that we might solve them, but simply to bring them to life, and to enable us to appreciate the wonder of the solution—if solution it be. Introductory courses in symbolic logic do not ask students to invent the subject. Almost never is a course organized around a major, open, unsolved problem—background knowledge and skills being acquired along the way, as a part of the effort to improve understanding of the problem and how it is to be tackled and perhaps, one day, solved.

The other course, introducing philosophy, I found less interesting. This was partly because I already had a background in philosophy, having earlier read works by G.E. Moore, Bertrand Russell, A.J. Ayer, and others, and having thought about philosophical problems for years, without quite realizing it. But—perhaps for this reason—it turned out I was rather good at philosophy. At last, something I could do!

As my first year at Manchester came to an end, I became nightmarishly obsessed with two philosophical problems: the mind/body problem, and Hume's problem of induction. I threw my mind into a torment in connection with the first of these problems: how could a mere brain, a conglomeration of neurons and synaptic junctions, however vast and intricately designed, give rise to consciousness, to inner experience, to thought, feelings, and perception, to our inner world? Was one to suppose that inner awareness arose as a kind of smoke from functioning neurons and synapses? Inner experiences, thoughts, and feelings seem to be intrinsically and utterly different, in their very nature, from any conceivable neurological process to be found in the brain: and yet it also seems absurd to hold, with Descartes, that our inner experiences, our thoughts and feelings, are utterly distinct from anything to be found in the brain, there being two distinct kinds of stuff in the universe, the mental and the physical.

But as if this was not bad enough, there was also Hume's problem to torment me: what possible reason can there be for holding that things will continue in the future more or less as they have done in the past? How can our knowledge of the present and past be known to have any relevance for the character of the future? We cannot know anything about the future until it is here, as the present, or departed into the past. At any moment, for all that we can ever conceivably know, anything whatsoever may happen. We cannot argue that in the past the future has resembled the past, and hence in the future too the future will resemble the past, because this presupposes just what is at issue, namely that the past is a reliable guide to the future. At any instant, for all we can ever know, a teaspoon may become an elephant, or a daffodil an ocean, and it is just as sensible to hold that the teaspoon will in the next second be an elephant as it is to hold it will continue to be a teaspoon.

I found myself caught in a nightmare of contradictory impulses. On the one hand, the whole problem posed by Hume was clearly absurd: there must be some simple way of refuting Hume decisively. But I could not see what it could be. On the other hand, here I was trying to refute Hume when what had so appealed to me about Hume's argument, when I had first learnt of it as a child, was that it is absolutely correct, and beautifully puts humanity's pretensions into proper perspective by demonstrating conclusively that the natural world must always be in a position to surprise and confound the scientific experts, however infallible they may claim their expertise to be. Hume's argument deserves to be affirmed and celebrated as providing liberation from the tyranny of expertise—and here was I, working against my instincts, furiously striving to demolish this wonderful and valuable argument as an absurdity.

The Riddle of Our Desires

My obsession with these two problems congealed into black despair. I arrived home for the summer vacation. My despair, my sense of inner blackness, excluded all thought of a holiday. I decided that I would write. It was now or never. I took a job working in a factory during the day, and in the evenings, in a state of terror, I began to write short stories in one notebook and, in a second one, I jotted down thoughts and feelings just as they occurred to me — writing whatever I wanted to write, free of the crushing burden of attempting to create literature. This latter activity transformed my life. "The riddle of the universe", I wrote, "is the riddle of our desires" (and twenty years later this became the central theme of *What's Wrong With Science?*, *From Knowledge to Wisdom*, and much of my subsequent work). What had I wanted? Pushed to the extreme of absurdity, I decided, to become God. My desire to discover the ultimate nature of the physical universe and reveal it to humanity so that everyone could know and understand the ultimate truth about our world, our existence — what was this but the desire for omniscience, for immortality, the desire to be God? Buried within the scientific enterprise, it seemed, there were these passionate, desperate, absurd strivings — to acquire God-like knowledge and understanding, God-like power, to become immortal, to become God. And likewise my desire to be a writer of genius, chart the hidden depths of the human heart, create worlds of experience more vivid and real than reality itself, disclose for everyone to see the supreme inner meaning and value of human life, the miracle of our existence — what was all this again but the desire to be God? In literature too, it seemed, there were intense, desperate, concealed longings for God-like status.

And all this was an awful mistake. In my desperate desire to be a genius, to be God, I had lost sight of something infinitely more precious: to be myself. It was not the impossibility of becoming God that struck home to me so forcefully, as I scribbled away in my diary of thoughts and feelings, but rather the appalling and grotesque undesirability of it. For me, infinitely more miraculous than being God was being myself — this unique and extraordinary being which only I could be. In striving to become a genius, God, I had been striving to destroy myself. I had never seen myself — such a small, humble, short-lived phenomenon in cosmic history — as anything worth being, for its own sake. Now it seemed infinitely precious. And I had the sense of a foetus inside me, my embryonic self, frozen and withered from long neglect, now just beginning to stir, to grow, to feel and see, utterly sensitive and naked to experience.

For the first time in my life I passionately desired to be myself. But what was I? I did not know. It was a mystery. This long neglected, hitherto despised I was a stranger to me. I had for so long trampled on myself in my desperate attempts to escape from myself into becoming an immortal genius that I now did not know what my poor, trampled self could be. At times I experienced terror as I felt myself ceasing to be.

I invented a theory. For the first months of our life, I decided, it must be that we do not know how to divide up what there is into "me" and "not me". There is simply "everything" — moving colours, sounds, feelings, pains and pleasures: a cosmos of experience. Then we discover how to separate out "that which is me" from "that which is not me": and we discover that we are a tiny, powerless being in a vast, all-powerful, largely unknown, sometimes terrifying universe. Dimly we remember a time when we were "everything", a time of blissful God-like status when the distinction between "me" and "cosmos" did not exist. In some way, so it seems to us, we have been disinherited of our rightful status in the scheme of things of being "everything". Without realizing what we are doing, we devote the rest of our life, in one way or other, to striving to attain again our original, proper status of being "everything". This is our due, our natural inheritance.

Two opposing strategies are adopted by people in their desperate struggles to become "everything". On the one hand people try to increase the size of the self, the minute "me", in the mad hope that eventually it will swallow up the entire universe — or at least as much of it as possible. Thus people strive to become more and more powerful, so that, by conquest, they come to dominate more and more territory, more and more people — until, eventually, pushed to the limit, the world itself quails before such all-commanding might. Others strive to increase the size of their identity by possessing more and more, enhancing the capacity to own by amassing wealth. Others seek to inherit the earth through their children — their progeny peopling the universe. Others again strive to become the universe by, quite literally, swallowing it up, obsessively and hopelessly eating and eating in an attempt to turn all that is "not me" into "me". And others seek to become the universe through science, through knowledge and understanding—the mind possessing, and even becoming, what is known and understood, the knower swallowing up and digesting even those vast cosmic tracts of space and time by knowing and understanding them.

On the other hand there are people who adopt just the opposite strategy: they seek to become "everything" by diminishing the self, the minute speck of "me", until it disappears altogether and only "everything" remains. This is the strategy of the mystic, who seeks the progressive annihilation of his self until it vanishes entirely, and there is only God: abrupt, ecstatic, devastating mystical union with God. It is the strategy, more generally, of all those Christians who strive to destroy their selfish self, strive to become humble and selfless, so that their will may become no more than the will of God, the self sunk into union with God by becoming a mere servant, a tool, a finger of God's purpose and presence in the world. It is the strategy of all those who endeavour to abase and annihilate their distinct identity before some vast "other", so that it becomes nothing but a part of, a servant of, the "other"— whether this be God, the Church, the Nation, the race, the future, the people, or whatever. Even those who seek oblivion in alcohol, in drugs, in trauma, in madness, the self-knowing self being obliterated beneath the "everything" of sheer sensation and experience, adopt a version of the second strategy. The strategy is adopted implicitly even by those suicides who hope that by destroying themselves they will become—everything: only the existence of the tiny self-knowing ego standing in the way, it seems, of the grandeur of becoming the cosmos.

These two strategies—to swallow everything up, and to be swallowed up by everything—on the face of it diametrically opposed, actually differ only in being different means to the same end: to become at one with God, with Nature, with Everything. Conquest and self-effacement, arrogance and humility, dominance and submission, selfishness and selflessness, apparent complete opposites, are actually but two sides of the same coin. And we all, helplessly, without quite knowing what we are doing, in our urgent hunger to find reality and fulfilment, throw ourselves into living out our own particular version of one or other of the two strategies for life. The life-goal of becoming at one with everything is of course the outcome of extending our actual life goals to the extremity of infinity and insanity. Most of us massively curtail such a goal in the light of what we deem to be possible in the given hard constraints of real life. Only a minority of us, the insane, the mystics, the saints, the Hitlers and Stalins, can live out in actual life the fantasy of being everything—everything of importance. Nevertheless our wildest dreams and longings, projected to infinity, even if dismissed as childish or mad, can still influence our actual goals in life, our actual life strategies. What we actually do is the achievable residue of our infinite hopes. In our dream life we devote ourselves to being at one with God; and in the constricting circumstances of our actual life this becomes: to be head of the firm; to become blind drunk yet again; to publish yet another scientific paper; to achieve promotion; to perform an act of selflessness, of self-abnegation. Viewed from this perspective, our life is bound to seem frustrated and absurd.

And all this—so my theory asserted—is tragically unnecessary, the outcome of an awful mistake. It all rests on a grotesquely mistaken view of the nature of the self, a mistaken view of the relationship between that which is "me" and that which is "not me". Influenced by Christian conceptions of the soul, and the Cartesian conception of the Mind, we are led to conceive of our identity as a bubble of mind stuff which floats precariously within a vast, impersonal physical universe. It is not just that we discover that we are not "everything", but only a tiny vulnerable body within an immense world that is not us: worse still, we discover that we are not even our own body— our "me" being no more than the intangible mind stuff of consciousness floating somehow within the interstices of the brain.

We are banished from the world, imprisoned within the bubble of our mind, all that we experience being no more than moving images within the bubble, caused by the utterly unknown, distinct physical world beyond, which will before long, and with complete indifference, crush us out of existence. Holding as we do, somewhere at the back of our mind, this nightmarish vision of being squashed up for life within the bubble of our mind, separated and excluded forever from the utterly distinct world of Reality which lies beyond, we take up our mad life project of becoming at least a part of Reality, either by trying to swallow it up, or by allowing it to swallow up us.

But this Cartesian picture of the relationship between "me", my consciousness, and "that which is not me", the physical universe, is entirely wrong. That which is within us, our inner conscious self, is just as unknown to us, just as much a mystery, as that which lies without. The stuff of our inner experiences is as real and as mysterious as the stuff of apple trees, stones, or sunlight. We are not this unknown inner world — anymore than we are the outer world surrounding our bodies. We are the outcome of the interaction between inner and outer worlds. We are as much the trees, the sky, the sounds of our footsteps scrunching on gravel, the action of walking in the world, as we are our experience of these things, located within our skull. Our identity is not made of mind stuff, utterly distinct and separate from the material world: rather our identity is the interaction between the world out there, and what lies within. Our identity is naked to the world. We become what we see, and hear, and touch, and do.

But as a result of conceiving ourselves to be utterly distinct from everything else, the easy flow of identity between what is within and without is disrupted. Desperate, hopeless attempts to become at one with Reality by trying either to swallow it up, or to get it to swallow up us, only make matters worse. The clenched muscles of our identity impose an even more restrictive barrier between our inner and outer worlds.

The crucial step is to recognize that there is a third way. In order to become ecstatically at one with Reality, to as great an extent as we please, all we need to do is to relax our clenched muscles of identity dividing off so artificially "me" from "not me", and we discover ourselves in the easy interplay between what is within and without. We will discover ourselves to be what we already are: a part, an aspect of, Reality. Strenuous, hysterical, and hopeless attempts to conquer or be conquered will fall by the wayside as we participate in the miraculous richness of Being.

This was my theory. It contains the seeds of the ideas of my subsequent work. In an emblematic but confused way, almost all the themes are there: emphasis on the need to call into question the aims of science and the aims of life, and the importance of relating the one to the other; the fundamental character of the problem "What is of value in life, and how is it to be realized?"; the need to change philosophy so that it takes up as its basic task to help us improve our solutions to this problem as we live; the importance of trying to understand human life as we enjoy and suffer it, imbued with meaning and value, as an integral part of the physical universe; the idea that experiential features are real features of things out there in the world, all views which deny this, from Cartesian dualism to physicalism, being wrong; the sense that being alive is a miracle — that which is of supreme value in existence lying in the rich particularity of our lives here on earth.

There is of course much to criticize in my "theory". Indeed, I subjected it to fierce criticism myself in ensuing years, and was able to develop my views as a result. But when I first enunciated it, in the summer of 1961, it came to me as a revelation, as a solution to the riddle of existence. Not only did I believe passionately in my theory; I lived it. What I thought of as my great discovery — that in order to realize what is of supreme value in existence we need to forego attempts to possess, or become possessed, and instead allow our self to emerge naturally as the interaction between unknown inner and outer worlds — this great "theory" of mine was but the intellectual husk of what I lived, what I experienced. In the space of a week or so, almost everything had changed. My black despair had gone. I found myself a new person in a new world, vivid, dramatic, sometimes terrifying. Now that for the first time in my life (so it seemed) I wanted to be myself, but did not know what this unknown, mysterious thing "myself" could be, during each day I found and lost myself a thousand times as I became and ceased to be what I saw, felt, heard, did, or became a part of. Every morning just before sunrise, I set off for a long walk through the beautiful Hampshire countryside in which my parents' house was situated. As the sun rose, it felt like the first day, the beginning of existence. Everything was indescribably fresh. I was newly created: and being myself, whatever it might be, seemed to be a wonder, something sacred. As I walked, I would lose myself in the changing perspectives of trees, hedges, hills, and sky: and the landscape would lose itself in me. I experienced the dissolution of the barriers between "me" and "not me", so that at times it seemed it was the landscape walking me through it, there being simply changing perspectives of landscape. And then I would run, frightened that I was about to dissolve away altogether, lose myself permanently to these trees, fields, and sky, and go mad. And throughout the day I would find and lose myself in the changing circumstances of my surroundings. In one of his letters John Keats remarks, "if a sparrow were before my window, I take part in its existence, and pick about the gravel". So it was with me. I would meet two friends for five-minute conversation, and I would become this meeting, this conversation. This would be everything, my whole identity. And when the meeting came to an end, I would experience the terror of dissolution of self, until I found a new self in what happened next. Whereas before I had been shut up in my solitary Cartesian prison, utterly excluded from the real world, surrounded by the impersonal, unknown physical universe, I was now released into a world rich in colour and drama, my identity as much in things around me as in my body or brain, passionately and helplessly becoming and ceasing to be the things I experienced. I can remember staring at a stalk of grass: never had there been such vividly green grass; and this was not some object remote from me about which I could only obtain distant, misleading clues through perception or touch; as I stared at it I became it, or it became me. I knew the stalk of grass from within itself.

And it was not just things that I met in this raw, absolute way; it happened with people too. Whereas before I had been locked away from others by my terror of being known and annihilated, I now plunged into communication with others, friends and strangers alike, with reckless, uncalled-for intensity, convinced that there could be nothing more important than that we should know each other without reservation while so briefly and miraculously alive. I sought intimacy swiftly and unselfconsciously, entirely unperturbed when this led to embarrassing or absurd consequences. On one occasion, in London, a friend took me back to the house he shared with others. We entered the living room together: my friend's fellow lodgers, all complete strangers to me, were watching television. I strode across the room and, without a moment's thought, turned the television off, quite sure that meeting each other was of infinitely greater significance than watching flickering images on a screen in silence.

Up till now I had instinctively presumed, without quite realizing it, that that which is of supreme value in existence must be something hidden and remote, buried deep in the structure of the cosmos, or in the intricacies of the human psyche. Now I experienced supreme value as something brazenly apparent in my immediate surroundings, as something I could see, touch, and become a part of. What hitherto I had only had a glimpse of in isolated, battering moments of ecstasy and terror, I now endured as a day by day reality, it being the new world in which I found myself. It was as if I had had my familiar self and world dissolved away by some psychoactive drug, and I now experienced reality naked and raw — except that there were no hallucinations, and the whole experience lasted not for a few hours but for six weeks.

One point still worried me: the megalomania of philosophy, in which I seemed to be caught. Scientific impulses to know and understand, and artistic impulses to create, might contain within them secret, unacknowledged desires to acquire God-like status: but in the case of philosophy, the desire to become God seemed blatant, horrifying, and grotesque. The great philosophers were, I felt, little better than would-be great dictators, who tried by intellectual means to establish absolute power over humanity forever. Each philosopher dreamed up his own personal vision of how he desired the universe, life, and society to be, and then sought to foist this personal vision onto the rest of us by arguing that reason alone proved the vision to be true, the one and only absolute, objective reality. The philosophical picture of reality is not put forward honestly as a personal wish or dream, as a suggestion or proposal, a possibility open for the rest of us to consider, to accept, reject, or modify as we please. It is put forward as final Truth established and authenticated by mighty Reason forever—Truth that we are all obliged to accept and adopt. By means of this trick of dressing up what he desires as the commands of Reason alone, the great philosopher seeks to hypnotize humanity intellectually, so that the rest of us come to believe, value, and do what he desires and dictates. Under the impression that we are observing the edicts of reason, we quietly become the great philosopher's slaves.

Plato—generally acknowledged to be one of the very greatest of philosophers—seems an especially blatant case in point. In *The Republic*, Socrates argues on behalf of Plato that society must be organized in the way the philosopher deems to be right, for he alone—via his intellectual perception of the entities of mathematics—has perceived the Form of the Good and thus is able to know what constitutes the good society. In brief, he, Plato, needs to be given absolute power for he, alone, knows what is good for the rest of us.[14]

---

[14] A few years later I read Karl Popper's *The Open Society and Its Enemies*, and was pleased to discover I was not alone in holding Plato to have had dictatorial aspirations. Bertrand Russell in his *History of Western Philosophy* comes to a similar judgment, as does Richard Crossman in his *Plato Today*.

Plato is perhaps an extreme case. I felt, however, that the Platonic lust for power, the Platonic urge to become God, concealed beneath a smokescreen of professed wisdom, was inherent in the very enterprise of philosophy itself, as traditionally conceived. For not only Plato but other philosophers, too, traditionally try to show that a personal vision of how things are and ought to be is the unique and absolute Truth, decreed by Reason, which ought to be accepted as such by humanity forever. What is this but the attempt to become the dictator of humanity by intellectual means?

All this might horrify me: but was not I also guilty of just such a dictatorial project? After my (admittedly highly anti-Platonic) mystical experience of reality, did not I now desire to tell humanity of my great discovery? For had not I discovered the solution to the great riddle of existence?

Recoiling in horror from this realization, I decided that in future we cannot possibly put our trust in the rare great philosopher or prophet: we must all become prophets. We must all make up our cosmos, our life, our world of value, for ourselves and each other. We are all philosophers, even as children. A vital part of the intellectual deception of the great philosophers —the would-be great dictators—is to fool us into thinking that philosophy is a highly abstruse field, much too difficult for most of us, our confidence in our capacity to think for ourselves thus being undermined. This is a crucial step in Plato's argument. Much subsequent education, right down to the present, conspires to make most of us lose rather than gain confidence in our capacity to think responsibly and seriously about fundamental issues for ourselves. And experts of all kinds—scientific, medical, technological, religious, academic, even political—seek to cloak their expertise in jargon so that it is incomprehensible to the layman, thus further undermining our confidence in our ability to judge and know for ourselves. This vast, elaborate conspiracy to deprive us of our wits, our capacity to know and understand our world, must be resisted. What we urgently need is a democracy of prophets.

But was not I still a kind of would-be intellectual dictator, in preaching this great message that we must all be prophets and make up our world as we live? For here was I, a kind of privileged meta-prophet, dictating to humanity the great Truth that we are all prophets.

It seemed to me that the only way this final dreadful charge of intellectual dictatorialism could be avoided was to abolish absolute Truth itself. There are stories, myths. The great mistake is to take any one story to be *the* story, *the* absolute Truth, Reality itself. It is the everlasting temptation to do this which creates the impossible problem of how to reconcile the physical universe with our human world as we experience it. We suppose that the scientific myth is the Truth, Reality itself: and then we are confronted with the impossible problem of accommodating all that science leaves out: sensory qualities of things, thoughts and feelings, meaning, freedom and value. The solution is to reject the initial premise. Science does not provide us with a privileged access to Truth and Reality: the scientific conception of the world is one myth amongst others, in some respects better than others, in other respects worse. Scientific entities like electrons and protons are fictional objects like gods: useful for certain purposes, but not to be taken too seriously.

I thus lapsed into a kind of relativism. Nevertheless, convinced that I had made discoveries of momentous importance — above all, that the riddle of existence is the riddle of our desires, philosophy being about the vital problem of what we should do with our lives — I was eager to return to Manchester to astound my philosophy mentors with the intellectual riches I had stumbled across. Academic philosophy was not the tedious discipline I had so far encountered, concerned only with sterile problems of dead knowledge: it was vital and alive, charged with personal experience, ready to grapple with urgent and basic problems of life.

Back at Manchester, I found I could not open my mouth. The dramatic and vital enterprise that philosophy had now become for me seemed to have no connection whatsoever with philosophy as conducted in the Department at Manchester

University. The idea that philosophy might have something to do with life, with the great mysteries of existence, with problems of living in the real world, seemed to be grotesquely out of context in the Philosophy Departmental Seminar at Manchester. Merely to take philosophy seriously seemed laughable. The rich and extraordinary world that I had discovered that summer began to fade away. My mouth was full of concrete, and I could not speak. I began to despair.

In the third year, things became really grim. The course became devoted to Oxford philosophy, which struck me as the absolute nadir of what philosophy might be. The proper task of philosophy is to articulate our most urgent, general, and fundamental problems — problems of thought and life — and propose and critically assess possible solutions, or at least help keep alive this vital activity. Oxford philosophy, and much analytic philosophy which stems from it, quietly denies, by implication, that any such activity is possible for philosophy at all. For it is implicitly — and idiotically — taken for granted that philosophy cannot be about problems that concern the real world, because philosophy is not empirically based. Hence, philosophy can do no more than analyse concepts. Not only does this deny to philosophy the very possibility of its proper task. It condemns philosophy to intellectual dishonesty. For the results of conceptual analysis are presented as being no more than conceptual clarification. But inherent in the meaning of words there lurk factual, metaphysical, value, and even political assumptions. Such assumptions must be implicit in the supposed "conceptual clarifications" of analytic philosophy — even though this will, of course, be denied. Thus Gilbert Ryle, in his *The Concept of Mind*, claims merely to analyse mental concepts. Actually, the book insinuates the doctrine of behaviourism, but this is done in a covert and dishonest fashion, and is explicitly disavowed.[15] The

---

[15] Behaviourism emerges merely as a result of the factual assumptions Ryle makes in discussing the "use" (and so meaning) of mental concepts. Different factual assumptions would lead to different results. Consider a

absolute pits of Oxford philosophy, for me, was J.L. Austin's *Sense and Sensibilia*, so smugly and idiotically reducing the problem of the nature of reality to the meaning of the word "real".[16]

I was so appalled by Oxford philosophy, and so frightened of becoming indoctrinated by such rubbish, that I stopped attending lectures entirely, but nevertheless managed to pass the final exam and obtain my degree.

MA Thesis, Karl Popper, and J.J.C. Smart

I decided to spend two years attempting to capture in an MA thesis what I had discovered in the summer of 1961, and then somehow mislaid in my despair at not being able to open my mouth about it. I still had this sense I had discovered something of profound significance, even though I now felt I no longer knew quite what it was. My thesis, I decided, should tackle the problem of how to reconcile the two worlds of my megalomaniacal youth: the universe of physics on the one hand, and the human world of common sense, of experience, consciousness, meaning, and value on the other. My initial idea was that these should be treated as two *myths*, two stories, neither to be taken too seriously as the one and only *truth*.

But then I discovered work by Karl Popper, and in particular his argument that we cannot verify scientific theories, we can only refute them. I was impressed. This was close to my own view of the matter, but also subtly different. Popper was a firm believer in truth, even though he also thought it was very difficult to get hold of. Physics, for Popper, very definitely, was not

---

society which believes some members are philosophical zombies, devoid of consciousness, while others are conscious because they possess Cartesian minds. The use of "conscious" in such a society would reveal that, in order to be conscious, one must possess a Cartesian mind. This just indicates how dishonest and spurious Ryle's whole procedure is.

[16] "What is the nature of reality?" is an authentic problem because we are confronted by a number of rival views about the nature of the world, and we want to know which is true—for example, naïve realism, physicalism, phenomenalism and, the one I favour, experiential physicalism.

one story amongst others, no more or no less valid than others. As a result of reading Popper, my whole view of the task of philosophy was transformed.

Up until then my approach to philosophy was to take a problem—"Do we see stars?", "Is the mind the brain?"—and write a kind of mini-drama, invisible protagonists deploying arguments for and against. In this way, I thought, one could acquire philosophical insight. For Popper, this was clearly no good at all. One should begin with a serious, open, difficult problem (excellent!) and one should try to solve it. One should try to get at the truth of the matter, even though this might be very difficult, and even though, even if one did solve the problem, one could have no assurance that one had definitively solved it. All our knowledge, all our attempted solutions, can only be, forever, guesses, conjectures. And one should find out about the history of attempts to solve the problem, assess critically past attempted solutions, and then attempt, if possible, to do better. Above all, one should write in as simple and clear a way as possible, avoiding all jargon and technicalities, unless they proved absolutely necessary.

This passage of Popper's in particular made a big impact:

> The belief of a liberal—the belief in the possibility of a rule of law, of equal justice, of fundamental rights, and a free society— can easily survive the recognition that judges are not omniscient and may make mistakes about facts and that, in practice, absolute justice is hardly ever realized in any particular case. But this belief in the possibility of a rule of law, of justice, and of freedom, can hardly survive the acceptance of an epistemology which teaches that there are no objective facts; not merely in this particular case, but in any other case: and that the judge cannot have made a factual mistake because he can no more be wrong about the facts than he can be right.[17]

---

[17] Popper (1963, p. 5).

I found this argument utterly convincing. It was actually profoundly immoral not to believe in objective factual truth, the world of objective fact, existing independently of us, whether we were aware of it or not. It was just this, I realized, ashamed of myself, that I had abandoned with my view that there are just stories, no one story having the right to claim to be *the* story, the one *true* story. The world exists, whatever stories we may make up about it. Many are false. Some parts of some are, no doubt, true, in that what they say to be the case, really *is* the case. Far from stories swallowing up the world, stories are just a tiny bit of the vast universe which exists entirely independently of what we think about it, apart from the tiny bit of the universe that happens to be ourselves.

I would have to revise entirely my hopelessly inadequate attempted solution to my problem. I decided that the worlds of (1) physics and (2) the experiential are not different *stories*, neither having the right to claim to be *true*. Rather, they are concerned with different *aspects* of all that there is. The problem now became to pin down exactly *what* aspect each is concerned with, and what, exactly, prohibits physics from saying anything about the experiential aspect.

I became an occasional student at the LSE, and attended Popper's lectures and seminars. I was immensely impressed with the originality and clarity of what he had to say. His incidental remarks seemed to me to demolish casually whole swathes of received views, public opinion, and prejudice. He spoke against the idea that the future will decide about the merit of works of art. He characterized a great deal of contemporary philosophy as amounting to no more than a "word salad": words are tossed about, but nothing substantial is achieved. He stressed, again and again, the importance of beginning with a statement of one's problem. He pointed out that critical rationalism is relevant even in the arts. An artist tries things out, criticizes what he has done, tries again, and so on until what has been done is deemed sufficiently good to survive.

I read Popper's *The Open Society and Its Enemies*, and quite literally wept with relief and joy. At last I had found a work of philosophy that had profound things to say about a profound problem: the severe difficulties that face the open, liberal society, even some of the most revered thinkers, above all Plato and Marx, being enemies of it. I was especially impressed with the way Popper was able to show that the rational society *is* the open society (highly relevant to ideas I was subsequently to develop). Given pre-Popperian conceptions of reason, one might suppose that the rational society would be a grim dictatorship of reason, rigidly controlled by rules of reason. Once one accepts Popper's critical rationalist conception of reason, however, it becomes clear that the rational society is a society in which criticism flourishes, which in turn requires toleration of diversity of views and values, which *is* the liberal, open society. There could not be a greater contrast with the spurious trivialities of Oxford philosophy. I remember thinking in one of his lectures: "This man is a great philosopher, an historical figure, standing in line with David Hume, Immanuel Kant, John Locke, and Rene Descartes."

Later, I came rather to pity Popper. He had had this profoundly important idea. Theories cannot be verified in science, they can only be falsified: nevertheless, it is this that makes scientific progress possible, and so incredibly successful. This process of subjecting theories to ferocious attempted falsification can be generalized so that, whatever we are doing, we can hope to make progress if we subject ideas—attempted solutions to problems—to ferocious criticism. I had no doubt about the importance of the idea—its widespread implications. And yet it seemed to be extraordinarily difficult to get the significance of the idea widely appreciated, not because the idea is esoteric and difficult to understand, but the exact reverse, because it is so simple. In those days, around 1964, few philosophers thought much of Popper. One or two famous scientists admired his work: Peter Medawar, Hermann Bondi, John Eccles. The public had never heard of him. I saw Popper condemned endlessly to repeat himself, endlessly and hopelessly condemned to trying to tell the world of the fundamental importance, for all that we do, of subjecting our imaginative ideas to severe critical scrutiny. Never, never, I vowed, must I allow myself to suffer a similar fate.

And at the time I did not think it was remotely likely that I would, because I had become convinced that Popper had essentially sorted out the fundamental problems of philosophy — the fundamental problems with which I had been so agonizingly obsessed. I heaved an immense sigh of relief. I no longer had the sense that I had discovered and lost something of profound significance for humanity which I must try to recover, articulate, and put into the public arena. Popper had done it — or would soon have done it, I reckoned, when people woke up to the significance of his work. I could relax. There were, it is true, a few questions Popper had not discussed, in particular my problem concerning physics and common sense, the experiential world. There was still something for me to do. But this was little more than pottering about on the sidelines. It did not concern the epoch-making matter of the future of humanity.

Back in Manchester, I discovered another excellent book: J.J.C. Smart's *Philosophy and Scientific Realism*. This defended *physicalism* — the doctrine that the universe and everything in it is made up exclusively of the fundamental entities of physics: electrons, protons, neutrons, and photons, or whatever these entities might be. The experiential world, thoughts and mental processes, aesthetic and moral qualities: all these are nothing but physical processes that are, ultimately, interactions between the basic entities of physics. Nothing but physics exists.

Not for one moment did I believe this. But I found Smart's book admirable, because of its clarity, its lucid exposition and defence of an important thesis, and its refusal to have anything to do with the sophisticated absurdities and dishonesty of Oxford analytic philosophy. In one bound, Smart had broken the idiotic convention that philosophy has nothing to do with the real world. He defended brilliantly a sweeping doctrine about the real world.

His book provided me with just the framework, the background, I needed for my thesis. I could devote much of my book to demonstrating what was wrong, or inadequate, in Smart's physicalist doctrine, and what one needed to put in its place. Physicalism, in my view, was absolutely correct as long as one did not claim, as Smart did, that everything comes within the scope of physicalism. It provides us with a picture of an aspect of what there is, but does not tell us about *all* that there is.

My task, as I have said, was to pin down exactly *what* aspect the physical, and the experiential, are concerned with, and what, exactly, prohibits physics from saying anything about the experiential aspect.

Physics, it occurred to me, seeks to *predict*. It is concerned only with that aspect of things which determines the way events, or isolated systems, evolve in time and space. As I wrote, something entirely unexpected and rather extraordinary occurred. Out of the tip of my biro, as I scribbled, had come, entirely unforeseen, a decisive refutation of one the pillars of modern philosophy, one of its strongest, best established doctrines. I refer to Hume's account of causation. Hume had argued that there could not be any kind of *necessary connection* between cause and effect. Nothing in one event, at one moment, could possibly determine, with necessity, what occurred at the next instant. I discovered that we must be even more epistemologically modest than Hume. For all we can know, necessary connections between successive states of affairs may well exist. Indeed, it is precisely the task of physics to try to discover what such a necessary connection may be. Theoretical physics seeks to discover *necessitating properties* of fundamental physical entities. If an electron is electrically charged then, *of necessity*, it will accelerate in a prescribed way when placed in an electric field. Indeed, all ordinary physical properties—solidity, rigidity, opacity, and so on—carry implications about how the object that possesses the property will, of necessity, *change*, or *resist* change, in certain circumstances.

Physics, then, is exclusively about what may be called the *causally efficacious* aspect of things, that aspect which everything has in common with everything else, and which determines *necessarily* (but perhaps probabilistically) how events unfold. Everything not required in order to predict how events unfold — the experiential aspect of things — will be ignored by physics. Given *any* isolated system — any physical system isolated from all influences from outside — physics seeks (in principle) to be able to predict how that system evolves, given a specification of its state at some instant. But physics is only interested in predicting specifications of the state of the system *when described in the same terms, in order to predict further states of the system*. Physics is thus not interested in predicting *everything* about the system. If it contains a conscious person who sees a yellow daffodil, has the visual experience of seeing the yellow daffodil, thinks "that's a daffodil", and then says "that's a daffodil", then physics will (in principle) describe the *physics* of all this: the molecular structure of the daffodil, the light reflected from the daffodil, the nerve impulses that travel up the optic nerve and around the brain of the person, the contraction of muscles, the sound waves emitted from the vocal chords and the mouth. But *what the daffodil looks like, what it is to experience seeing the yellow daffodil, what it is to think "this is a daffodil"*, and what "this is a daffodil" *means*, all this is ignored by physics because ignoring it does not in any way undermine its predictive task.

But it went further. It was not just that physics did not need to specify experiential qualities to achieve its predictive task successfully. I discovered an argument which showed conclusively that physics could not predict the experiential features of things, even if it wanted to (as it were). All physical statements — all physical concepts — are such that you do not need to have had any special kind of experience in order to understand them. Being blind from birth does not debar you from understanding the whole of optics, or the wave theory of light, just as well as any sighted person. But when it comes to experiential features, all this is dramatically different. If you are to know what "yellow" means, or what it is to assert "that daffodil is yellow", then you do, at some time in your life, need to have experienced the visual sensation of yellowness. A person blind from birth is not thereby debarred from understanding any part of physics, but he is debarred from knowing what it is that "This daffodil is yellow" asserts. This means that no conglomeration of statements of *physics* can ever imply a statement such as "This daffodil is yellow" where "yellow" is understood to refer to the perceptual or experiential quality, what we normally sighted people see when we see daffodils.

The silence of physics about colours, sounds, smells, tactile qualities as we experience them does not mean that these experiential qualities do not really exist out there in the world around us. All it means is that these qualities are just the sort of properties that physics does not need to specify, predict, or describe in order to fulfil its predictive, explanatory task. And furthermore, even if physics wished to predict these experiential qualities, it could not, because physics is such that no special kind of experience is required to understand it, but experiential qualities are such that special kinds of experiences *are* required to understand them, know what they are.

And precisely the same considerations arise in connection with our inner experiences. The silence of physics about our inner experiences, our thoughts and feelings, does not mean that these mental features do not really exist. All it means is that these mental features of brain processes are just the sort of features that physics does not need to specify, predict, or describe in order to fulfil its predictive, explanatory task. And furthermore, even if physics wished to predict these mental features of our brain processes, it could not, because physics can be understood without one having to have any special kind of experience, whereas the mental features of brain processes are such that you do have to have, in your own brain, just those brain processes in order to know what the mental features of those brain processes are.

I had solved my problem of specifying precisely what *aspect* of things physics is concerned with, and what *aspect* of things common sense, or the human world view, is concerned with. I had solved the problem of understanding how our human world, full of experiential qualities, could exist embedded in the physical universe.

And I had done more. Not only had I refuted Hume on causation. I had solved the philosophical part of the mind/brain problem. I had demonstrated, indeed, that the mind/brain problem is, in important respects, similar to the "yellow daffodil/daffodil as physical object" problem.[18] (Few philosophers today appreciate that the so-called hard problem of consciousness was solved long ago in 1965, the year I finished my MA thesis.)

All this gave precision to the explosive discovery I had made in the summer of 1961 that Cartesian dualism is wrong, mental or experiential features exist in the world external to us as well as within us, that we are, in a sense, as much a part of the world external to us as within us, and our inner world is as mysterious to us as the world external to us.

---

[18] Some years later, I added an important item to the argument. In opposition to the claim that physics cannot predict experiential features, it could be argued that physics could be extended so that it does predict these features. All that needs to be done is to add to physics extra postulates correlating physical features with experiential ones. Such an extended physics would, however, I subsequently argued, be hopelessly non-explanatory. Each extra postulate would be horrendously complex, and there would also be a vast number of them, to include the vast number of different kinds of possible experiential features and inner experiences that there are. Physical theory extended in this fashion would become hopelessly complex, *ad hoc,* and non-explanatory. The simplicity and unity—and thus the explanatory character—of physical theory would be wholly sabotaged. As I put it decades later, "There is, in short, an *explanation* as to why physics does not, and cannot, include the mental, the experiential. If it did, the extraordinary explanatory power of physical theory would vanish. Excluding the experiential is the price we pay for having the marvellously explanatory theories that we do have in physics": see Maxwell (2011a), first published online in 2003 at http://philsci-archive.pitt.edu/archive/00002238/.

Some years later, I argued that not only experiential features but value features too exist in the world around us, the latter perceived (fallibly) by our emotional responses to things.[19]

I wrote up my thesis, obtained my MA, and published three papers, in 1966 and 1968, which spelled out what I had discovered.[20] I assumed that these papers would have an explosive impact on philosophy. Not at all. There was absolute silence.

However, eight years later Thomas Nagel published 'What is it Like to Be a Bat?', and twenty years later Frank Jackson published 'What Mary Didn't Know'.[21] These papers did no more than express, perhaps in a somewhat more vivid way, a fragment of what I had argued for in 1966 and 1968.[22] Unlike my papers, however, they had an immense impact, not only on philosophy, but on psychology, cognitive science, artificial intelligence, and neuroscience as well. Decades later, in 1999, I wrote to Nagel and Jackson to ask them if they had come across my papers—and I sent copies. Nagel replied with great generosity, "There is no justice. No, I was unaware of your papers, which made the central point before anyone else". Jackson admitted he had read my "Understanding Sensations". Jackson echoed my argument in his paper without acknowledgement. Some time later I met Jackson and raised the matter with him. He said he had now abandoned what tends to be called "the knowledge argument". So, having taken credit for my work, he had now repudiated it!

As for my Hume paper, some years later Fred Dretske, Michael Tooley, and David Armstrong[23] published work along somewhat similar lines, again without any reference to, or apparent awareness of, my much earlier paper.

---

[19] Maxwell (1976a, pp. 139–46; 1984 or 2007a, ch. 10; 1999; 2001, ch. 2; 2010a, ch. 4).
[20] Maxwell (1966; 1968a,b).
[21] Nagel (1974); Jackson (1986).
[22] See especially Maxwell (1966, pp. 303–8; and 1968b, pp. 127 and 134–7).
[23] Dretske (1977); Tooley (1977), Armstrong (1978; 1983).

Does any of this matter, apart from my personal pique at not getting recognition for my work? At the time, I was unconcerned, but now, in retrospect, I am not so sure. In both cases, only a bit of what I had argued for came to the attention of philosophers. Most of what I argued for is still ignored by, and unknown to, most philosophers. What I have done since in developing these ideas has been ignored as well.[24] As for my Hume paper, subsequent work by others suffers from failing to reproduce key points of my argument. In 1989, Bas van Fraassen criticized anti-Humean accounts of physics on the grounds that if physical laws are necessary, they cannot be empirical, if empirical they cannot be necessary.[25] This criticism is lethal against the views of Dretske, Tooley, and Armstrong, but does not apply to my 1968 paper, which van Fraassen does not mention. In my paper, I stressed that physical theories need to be interpreted so that physical laws are analytic, and so necessary, all the factual content of the theory being concentrated in the assertion: such and such entities exist with such and such necessitating properties.

From my own personal standpoint, by far the worst consequence of the neglect of my first three publications was that when, a few years later, I really did have something important to communicate, concerning the future welfare of humanity, I failed. If my early work had received the recognition that was its due, later far more important work would undoubtedly have won far greater attention. I would not have had to struggle for decades to get it noticed, and still fail, even today.

London and the Comprehensibility of the Universe

After obtaining my MA, I taught philosophy of science in the Philosophy Department at Manchester for a year. Then, in 1966,

---

[24] See my (1984 or 2007a, ch. 8, replies to objections 6 and 7, and chs. 9 and 10; 2000b; 2001; 2009a; 2010a; 2011a; 2010b, pp. 677–84). See also note 19 on page 159.
[25] van Fraassen (1989).

I got a job teaching philosophy of science in the Department of History and Philosophy of Science at University College London, where I remained until my early retirement in 1994.

In London, I took up three projects: (1) to write a book expounding my solution to the human world/physical universe problem (encapsulated in my MA thesis and first three papers); (2) to develop a fully micro-realistic, fundamentally probabilistic version of quantum theory; and (3) to assess the validity of criticisms then being made of Popper's views about science. The first project fell by the wayside.[26] The second led to a long sequence of papers which did result, finally, in a micro-realistic, probabilistic version of quantum theory free of the defects of the orthodox version, able to solve the quantum wave/particle problem, and empirically distinct from the orthodox version (although not yet, as far as I know, put to the test of experiment).[27] The third project led to my profound discovery — or

---

[26] For references to my subsequent work on this problem see last but one footnote.

[27] See Maxwell, "A New Look at the Quantum Mechanical Problem of Measurement", *American Journal of Physics 40*, 1972, pp. 1431–5; "Alpha Particle Emission and the Orthodox Interpretation of Quantum Mechanics", *Physics Letters 43A*, 1973, pp. 29–30; "The Problem of Measurement — Real or Imaginary?", *American Journal of Physics 41*, 1973, pp. 1022–5; "Towards a Micro Realistic Version of Quantum Mechanics", *Foundations of Physics 6*, 1976, pp. 275–92 and 661–76; "Instead of Particles and Fields", *Foundations of Physics 12*, 1982, 607–31; "Quantum Propensiton Theory: A Testable Resolution of the Wave/Particle Dilemma", *British Journal for the Philosophy of Science 39*, 1988, pp. 1–50; "Beyond Fapp: Three Approaches to Improving Orthodox Quantum Theory and An Experimental Test", in *Bell's Theorem and the Foundations of Modern Physics*, edited by A. van der Merwe, F. Selleri and G. Tarozzi, World Scientific, 1993, pp. 362–70; "Particle Creation as the Quantum Condition for Probabilistic Events to Occur", *Physics Letters A 187*, 1994, pp. 351–5; "A Philosopher Struggles to Understand Quantum Theory: Particle Creation and Wavepacket Reduction", in *Fundamental Problems in Quantum Physics*, edited by M. Ferrero and A. van der Merwe, Kluwer Academic, London, 1995, pp. 205–14; "Does Probabilism Solve the Great Quantum Mystery?", *Theoria vol. 19/3*, no. 51, 2004, pp. 321–36; "Is the Quantum World Composed of Propensitons?", in *Probabilities, Causes*

what I cannot help but see as a "profound discovery": the urgent need to transform academia so that it gives intellectual priority to our problems of living, and devotes reason to the task of helping humanity make progress towards as good a world as possible.

Aware of the criticisms swirling around Popper's work, in part because of a review I wrote for *Nature* of a book comparing and contrasting Kuhn and Popper,[28] I decided to write a paper in which I would work out, for myself, whether any of these criticisms really struck home.

According to Popper, as everyone knows, science cannot verify theories, but can only refute them. This sounds very negative, but actually it is not, for science succeeds in making such astonishing progress by subjecting its theories to sustained, ferocious attempted falsification. Every time a scientific theory is refuted by experiment or observation, scientists are forced to try to think up something better, and it is this, according to Popper, which drives science forward. Thus, in order to maximize the chances of progress, scientists should put forward theories that have as much empirical content as possible, and are thus as vulnerable to empirical falsification as possible — these theories then being subjected to severe attempted empirical falsification. The best that we can hope for from science is theories that have survived such a ferocious empirical onslaught.

I entirely endorsed Popper's point that we cannot verify scientific theories; we can only falsify them. I had grasped this point, in essence, as a child when I first learnt of Hume's arguments concerning induction and causation. Criticisms of Popper made by Kuhn, Feyerabend, Lakatos, and others did not seem to me to be too serious.[29] What did seem to me to be serious, however, was Popper's failure to do justice to the fact that

---

and Propensities in Physics, edited by M. Suárez, Springer, Dordrecht, 2011, pp. 219–41.

[28] See Maxwell (1971).

[29] See my (1972a).

science only ever accepts theories that are *unified*[30] or *explanatory*. As I have indicated, Popper holds that that theory should be accepted which has the highest empirical content and has survived the severest onslaught of attempted refutation, thus being best corroborated. Given an accepted *unified* theory, T, we can always formulate a rival, T*, which has higher empirical content, and is better corroborated, by adding onto T independently testable, tested, and corroborated conjectures.[31] In general, however, T* would be horribly *disunified* and, quite properly, would not be considered in scientific practice for just that reason. Popper's methodology thus persistently recommends acceptance of theories that would never be considered in scientific practice for a moment. His methodology is refuted.

Popper tried to overcome this difficulty by arguing that a high degree of simplicity (or unity) is the same as high empirical content. But what the above indicates is that this is not the case. We can increase empirical content and at the same time drastically *decrease* simplicity, or unity.

Subsequently, Popper proposed that an acceptable theory "should proceed from some *simple, new, and powerful, unifying idea*".[32] This does much better justice to scientific practice. But this "requirement of simplicity", as Popper calls it, contradicts the rest of his methodology. It contradicts the requirement that we should accept that theory which has the greatest empirical content and is best corroborated — which, as the above argument shows, can invariably be concocted to be a T*-type theory, horribly *complex* and *disunified*.

Quite properly, then, scientists only accept *unified, explanatory* theories even though better corroborated *disunified, nonexplanatory* rivals can always be formulated. But Popper cannot provide a rationale for this procedure. He cannot explain why

---

[30] See note 7 and associated text on page 119.
[31] We can also add to T* corroborated empirical laws which T cannot (yet) predict because the equations of T cannot be solved.
[32] Popper (1963, p. 241).

this procedure gives us the best hope of achieving scientific progress. Such a procedure is only a sensible one to adopt, so it would seem, if we have good reasons to accept the *metaphysical* thesis that the universe is such that a unified pattern of law governs phenomena. If the universe is not like that, and basic laws are *disunified*, then adopting the procedure will block progress. Popper, however, cannot appeal to such a metaphysical thesis. He excludes metaphysics from science.

Then it dawned on me—and it was a definite moment when I had this revolutionary idea: the only way to make sense of science is to see the whole enterprise as accepting, as a basic item of (conjectural) scientific knowledge, that the universe is such that there is some kind of unified pattern of physical law running through all phenomena, the universe being, in this sense, physically comprehensible.

In one respect, Popper's conception of science is highly unorthodox: all scientific knowledge is conjectural; theories are falsified but cannot be verified. But in other respects, Popper's conception of science is highly orthodox. For Popper, as for most scientists and philosophers, the basic aim of science is knowledge of truth, the basic method being to assess theories with respect to evidence, *nothing being accepted as a part of scientific knowledge independently of evidence*. This orthodox view —referred to above as *standard empiricism*—is, I realized, *false*. The fact that physicists only ever accept *unified* theories even though endlessly many empirically more successful *disunified* rivals can always be concocted means that science makes a big, permanent, and highly problematic assumption about the nature of the universe independently of empirical considerations and even, in a sense, in violation of empirical considerations—namely, at the very least, that the universe is such that all grossly *disunified* theories are false. Without some such presupposition as this, the whole empirical method of science breaks down.

Suppose physicists only ever accepted theories that postulate atoms, even though many empirically better corroborated theories are available which postulate other entities, such as fields. It would be clear that physicists thereby make the assumption that the universe is made up of atoms, whether this is acknowledged or not. Just the same holds in connection with physicists' persistent acceptance of *unified* theories, even though empirically better corroborated *disunified* rivals are available. Physicists thereby make the assumption that there is some kind of underlying unity in nature, whether they acknowledge this or not.

Popper, along with most scientists and philosophers, had misidentified the basic aim of science. This is not truth *per se*. It is rather truth *presupposed to be unified*, presupposed to be explanatory or comprehensible (unified theories being *explanatory*). Inherent in the aim of science there is the metaphysical — that is, untestable — assumption that there is some kind of underlying *unity* in nature. The universe is, in some way, physically comprehensible.

But this assumption is profoundly problematic. We do not *know* that the universe is comprehensible. This is highly *conjectural* scientific knowledge. Even if it is comprehensible, almost certainly it is not comprehensible in the way science presupposes it is today. For good Popperian reasons, this metaphysical assumption must be made explicit within science and subjected to sustained *criticism*, as an integral part of science, in an attempt to improve it.

The outcome is the *aim-oriented empiricist* conception of science I have indicated above.[33] This facilitates progressive improvement of problematic assumptions inherent in aims by representing aims, and associated methods, in the form of a hierarchy, it becoming possible to improve the most problematic aims low down in the hierarchy in the light of improving knowledge, and less problematic aims high up in the hierarchy.[34]

---

[33] See note 8 and preceding text on page 121.
[34] My criticisms of Popper and standard empiricism, and my arguments for aim-oriented empiricism, are to be found in "A Critique of Popper's

At first I thought that this aim-oriented empiricist view is in such flagrant contradiction with orthodoxy that no scientist would have upheld it. But then it occurred to me that Einstein in his later years had held a similar view. More important, I discovered that, in creating special and general relativity, he had successfully put into practice the methodology of discovery that becomes available once aim-oriented empiricism is adopted.[35] Science implements aim-oriented empiricism in practice, but this is obscured and obstructed by widespread acceptance of standard empiricism.

It then dawned on me that the aim of seeking *explanatory truth* is a special case of the more general aim of seeking *valuable truth* — of value for its own sake, or for practical ends. And this is sought in order that it be *used* by people to enrich their lives. In other words, in addition to metaphysical assumptions inherent in the aims of science there are *value* assumptions, and *political* assumptions, assumptions about how science should be used in life. These are, if anything, even more problematic than metaphysical assumptions. Here, too, assumptions need to be made explicit and critically assessed, as an integral part of

---

Views on Scientific Method", *Philosophy of Science 39*, 1972, pp. 131–52; "The Rationality of Scientific Discovery", *Philosophy of Science 41*, 1974, pp. 123–53 and 247–95; "Induction and Scientific Realism", *British Journal for the Philosophy of Science 44*, 1993, pp. 61–79, 81–101, and 275–305; "The Need for a Revolution in the Philosophy of Science", *Journal for General Philosophy of Science 33*, 2002, pp. 381–408; "A Priori Conjectural Knowledge in Physics", in *What Place for the A Priori?*, edited by Michael Shaffer and Michael Veber, Open Court, Chicago, 2011, pp. 211–40; "Popper's Paradoxical Pursuit of Natural Philosophy", in *Cambridge Companion to Popper*, edited by J. Shearmur and G. Stokes, Cambridge University Press, Cambridge, forthcoming, and works referred to in note 8. Aim-oriented empiricism has the great advantage over standard empiricism in providing solutions to fundamental problems in the philosophy of science — problems of induction, simplicity (or unity), verisimilitude, and the character of scientific method: see note 8 on page 121, and "Practical Certainty and Cosmological Conjectures", in *Gibt es sicheres Wissen?*, edited by Michael Rahnfeld, Leipziger Universitätsverlag, 2006, pp. 44–59.

[35] See my (1993a, pp. 275–305).

science, by scientists and non-scientists alike, in an attempt to improve them.

Released from the crippling constraints of standard empiricism, science would, I felt, burst out into a wonderful new life, realizing its full potential, responding fully both to our sense of wonder and to human suffering, becoming both more rigorous and of greater human value (intellectually and practically).

My Apparent Great Discovery

Then one day, walking back home from work, I stumbled across my (apparent) great discovery.

I had been immensely impressed by the way Popper had generalized his falsificationist conception of scientific method to form a notion of rationality, *critical rationalism*, applicable to all aspects of human life. Falsification becomes the more general idea of *criticism*. Just as scientists make progress by subjecting their theories to sustained attempted empirical falsification, so too all of us, whatever we may be doing, can best hope to achieve progress by subjecting relevant ideas to sustained, severe *criticism*. By subjecting our attempts at solving our problems to criticism, we give ourselves the best hope of discovering (when relevant) that our attempted solutions are inadequate or fail, and we are thus compelled to try to think up something better. By means of judicious use of criticism, in personal, social, and political life, we may be able to achieve, in life, progressive success somewhat like the progressive success achieved by science. We can, in this way, in short, learn from scientific progress how to make personal and social progress in life. This, for me, in a nutshell, was Popper's great achievement: to have come up with a revolutionary conception of the progress-achieving methods of science which he then went on to show had, when generalized, profoundly fruitful implications for a wide range of human endeavours.[36]

---

[36] This is the line of argument that takes one from Popper's *The Logic of Scientific Discovery* to his *The Open Society and Its Enemies* and *Conjectures*

It suddenly occurred to me: I could pursue a path parallel to Popper's. Just as Popper had generalized falsificationism to form critical rationalism, so I could generalize my aim-oriented empiricist conception of scientific method to form an aim-oriented conception of rationality, potentially fruitfully applicable to all that we do, to all spheres of human life. But the great difference would be this. I would be starting out from a conception of science—of scientific method—that enormously improves on Popper's notion. In generalizing this, to form a general idea of progress-achieving rationality, I would be creating an idea of immense power and fruitfulness.

I knew already that the line of argument developed by Popper, from falsificationism to critical rationalism, was of profound importance for our whole culture and social order, and had far-reaching implications and application for science, art and art criticism, literature, music, academic inquiry quite generally, politics, law, morality, economics, psychoanalytic theory, evolution, education, history—for almost all aspects of human life and culture.[37] The analogous line of argument I was developing, from aim-oriented empiricism to aim-oriented rationalism, would have even more fruitful implications and applications for all these fields, starting as it did from a much improved initial conception of the progress-achieving methods of science.

The key point is extremely simple—as I have already indicated. It is not just in science that aims are profoundly problematic. This is true in life as well. We all die. This in itself makes our life aims problematic. Government, industry, agri-

---

and Refutations. It has not, I think, received the attention that it deserves, partly, perhaps, because it tends to be stated in footnotes and asides, partly because it straddles too wide a stretch of disciplines for today's specialists to be able to take it on board. It is, however, in my view, Popper's great achievement.

[37] For some of these implications and applications see Popper (1976). I give a brief, informal account of how wisdom-inquiry can be seen as a development and improvement of Popper's philosophy in Maxwell (2009d, Preface).

culture, medicine, the military, diplomacy, business, education, the law: all have problematic aims. Above all, the aim of creating a good world is inherently problematic, for all sorts of more or less obvious reasons. Furthermore, it is not just in science that problematic aims are misconstrued or "repressed"; this happens all too often in life too, both at the level of individuals and at the institutional or social level as well. There is an urgent need—I began to realize—for science to acknowledge, openly and honestly, its real and highly problematic aims so that it could begin to put aim-oriented empiricism explicitly into practice, and thus explicitly improve its aims and methods as it proceeds. Science needs to do this, thus making explicit and apparent its at present implicit and covert exploitation of aim-oriented empiricism, so that we can all see clearly *what* this aim-improving meta-methodology is, and just how extraordinarily successful it is, as far as science itself is concerned. Science thus becomes a methodological paradigm, a methodological resource, for the rest of life. The task then becomes to feed aim-oriented rationalism, generalized from the methods of science, into personal, institutional, and social life, so that we may improve our aims and methods as we live—so essential if we are to realize what is genuinely of value to us in life.

But this task struck me as profoundly difficult to perform. Despite its immense desirability, its fundamental importance for the future of humanity, it would meet fierce resistance at all levels; personal, institutional, social, global. It occurred to me that the social sciences and humanities would need to take up, as a long term project, to work out how we might feed aim-oriented rationality into personal, institutional, and social life. This would require the social sciences to be pursued as social *methodology*, or social *philosophy*, fundamentally concerned to help us improve aims and methods in life rather than acquire knowledge about social phenomena. I discovered that Popper, in arguing from scientific method to rationality and the open society, had been anticipated by the *philosophes* of the French Enlightenment, and I read Peter Gay's great book *The Enlightenment: An Interpretation*, enthralled. But even though the *philo-*

sophes had had their hearts in the right place, they had blundered, I realized, in thinking the task was to develop social *science* alongside natural science. If the basic Enlightenment idea is to learn from scientific progress how to achieve social progress towards an enlightened world, and the task of social inquiry is to work out how to do this, then social inquiry needs to be developed as social methodology, helping us to get into social life progress-achieving methods generalized from those that have been exploited with such success in science.

It began to dawn on me that academia as a whole, in thrall to the idea that first, knowledge must be acquired so that, then, subsequently, it can be applied to help solve social problems, is irrational in a far more elementary, wholesale, and damaging way than I had realized. Inspired again by Popper, it seemed to me obvious that absolutely elementary rules of rational problem solving are that one (1) articulates, and seeks to improve the articulating of, the problem to be solved, and then (2) proposes and critically assesses possible solutions. If we take seriously the idea that the basic task of academia is to help us realize what is of value to us in life by educational and intellectual means then, at the most fundamental level, the task must be (1) to articulate problems of living, and (2) propose and critically assess possible solutions—possible actions, policies, political programmes, institutional innovations, philosophies of life. What enables us to achieve what is of value is what we *do*, or refrain from doing, not what we know. Even when new knowledge is required, as it is in medicine or agriculture for example, it is always what this enables us to do that enables us to achieve what is of value (except when knowledge is itself of value).

If it is to devote reason to the best interests of humanity, then academia, I began to realize, would need to be quite fundamentally reorganized. Social inquiry and the humanities would need to be at the heart of academia, promoting cooperatively rational resolving of conflicts and problems of living in the real world (as well as helping us improve aims and methods in life). This fundamental intellectual activity would influence aims and priorities of research in the natural and technological sciences, and would itself, of course, be influenced by the results of such research.[38]

But what really matters, I realized, is the thinking we engage in as we live, guiding our actions. A basic task for academia is to help us improve this vital, socially active thinking.

I was aware that I had rediscovered my great explosive idea of the summer of 1961: philosophy should be about life; the riddle of the universe is the riddle of our desires. But my initial idea had been radically transformed. It was no longer just philosophy which should be concerned with our problems of living, but the whole academic enterprise. "The riddle of our desires" had become "the profoundly problematic character of our fundamental aims in life, both personal and institutional, including even the aims of science". Or it had become, perhaps: the riddle of the desirable — the riddle of what is ultimately of value in existence. The outcome of generalizing aim-oriented empiricism to form a general conception of rationality, aim-oriented rationality, and then applying this to the task of creating a better world was an entirely new conception, not just of science, but of academic inquiry, with implications for all of life.

Every branch and aspect of academic inquiry needs to change, I realized, if it is to be what it is supposed to be: rationally organized and devoted to helping humanity achieve what is of value in life.

I was confronted by five revolutions. First, a revolution in the philosophy of science, from standard to aim-oriented empiricism. Second, a revolution in science itself, so that it comes to put aim-oriented empiricism explicitly into scientific practice. Third, a revolution in social inquiry and the humanities, so that they come to give intellectual priority to

---

[38] For a list of changes that need to be made if knowledge-inquiry is to become wisdom-inquiry, see Maxwell (2014a, pp. 55–60). See also http://www.ucl.ac.uk/from-knowledge-to-wisdom/whatneedsto change.

problems of living, themselves put aim-oriented rationality into practice and take, as a basic, long term task, to help humanity feed aim-oriented rationality into the social world. Fourth, a revolution in academia as a whole, so that it takes up its proper task of helping humanity realize what is of value in life. And fifth, and finally, the revolution that really matters: transforming the human world so that it puts cooperative problem-solving rationality and aim-oriented rationality into practice in life, so that we may all realize what is of value as we live insofar as this is possible.

At some point it occurred to me that all this had a devastating implication for my own personal life: I would have to take up my own portion of personal responsibility for the state of the planet. For each one of us, it is extraordinarily difficult to feel that the future of the world has anything to do with what we do personally, in our own life. I am just one among billions. Anything I do can only have a minuscule impact on the state of the planet. But this is of course true of all those other billions of people as well. Each one of us is in the position of being powerless before the juggernaut of history, and yet that juggernaut is composed of us, of our actions. We are all responsible, and yet, individually, have only the minutest of impact on the whole. It is only when a majority of us do begin to take some personal responsibility, in our billions of individual, personal lives, for our common future, that we human beings can hope to begin to shape our destiny together, to suit our own best interests, instead of suffering the consequences of billions of us living without concern for what the net impact of our billions of lives may be. We all need, it seemed to me, to put something like 5% of our life effort into a concern for the state of the planet—the wealthy and powerful, of course, able to do rather more than the poor and powerless. And all this applied directly to me. I could not escape. My philosophy was no longer merely an idea and an argument. It is for life. In particular: my life.

I wrote a book, *The Aims of Science*, in which I spelled out my discoveries. It was rejected by publisher after publisher. I became a bit demented, holding forth to friends and strangers alike on the need to transform our schools and universities. Then a friend told me of a friend of his prepared to publish a book by me expounding my ideas. I thought about it, and then wrote my first book, *What's Wrong With Science?*, in three weeks to meet the publisher's deadline. Most of it takes the form of a fierce debate about the issues between a scientist and a philosopher. No one convinces anyone of anything (although I hoped the reader would find the philosopher's arguments utterly compelling). I thought this book would release my idea into the world, but the publisher failed to understand that review copies had to be sent out, the book received only three reviews, and was in general ignored.

Then Basil Blackwell's agreed to publish a new book. Slowly and painfully, I struggled to put clearly into words the new universe of ideas, arguments, and values that I felt I had stumbled across. My argument was that the basic aim of academic inquiry should be not just to acquire knowledge but rather to enhance our capacity to realize what is of value in life. But what is this "capacity"? Very much as an afterthought, it occurred to me that as good a word as any is "wisdom" (even though this had, for me, all sorts of undesirable connotations). Thus was born the title of my second book: *From Knowledge to Wisdom*. This was published in 1984. It received critical reviews from philosophers, some of whom criticized me for defending doctrines I explicitly *rejected* in the book! It received a supportive review from Mary Midgley,[39] however, and a glowing review in *Nature* by Christopher Longuet-Higgins, who wrote:

> Maxwell is advocating nothing less than a revolution (based on reason, not on religious or Marxist doctrine) in our intellectual goals and methods of inquiry... There are altogether too many

---

[39] She wrote "a strong effort is needed if one is to stand back and clearly state the objections to the whole enormous tangle of misconceptions which surround the notion of science to-day. Maxwell has made that effort in this powerful, profound and important book": see Midgley (1986).

symptoms of malaise in our science-based society for Nicholas Maxwell's diagnosis to be ignored.[40]

Unfortunately, my diagnosis has been ignored, and that has something to do with the troubles we face today. We strive to achieve economic growth, more industry and agriculture, more wealth, longer lives, more development, housing, and roads, more travel, more cars and aeroplanes, more energy production and use, greater security by means of greater military might. These things seem inherently desirable and many are, in many ways, highly desirable. But our successes in achieving these ends also bring about global warming, war, vast inequalities across the globe, destruction of habitats and rapid extinction of species, depletion of finite natural resources such as oil, pollution of earth, sea, and air—even the credit crunch of 2008 and global recession. All our current global problems are the almost inevitable outcome of our long term failure to put aim-oriented rationality into practice in life, so that we actively seek to discover problems associated with long term aims inherent in our current endeavours, actively explore ways in which problematic aims can be modified in less problematic directions, and at the same time develop the social, the political, economic, and industrial *muscle* able to change what we do, how we live, so that our aims become less problematic, less destructive in both the short and long term. We have failed even to appreciate the fundamental need to improve aims and methods as the decades go by. We have failed to see this even in the case of science. Our very ideals of rationality are such that they fail to help improve aims. Conventional ideas about rationality are all about *means*, not about *ends*, and are not designed to help us *improve* our ends as we proceed. Implementing aim-oriented rationality is essential if we are to survive in the long term, but academia does nothing to promote this idea, and has failed, so far, even to entertain the idea.

---

[40] Longuet-Higgins (1984).

Einstein put his finger on what is wrong when he said: "Perfection of means and confusion of goals seems, to my opinion, to characterize our age."[41] This outcome is inevitable if we restrict rationality to *means*, and fail to demand that rationality — the authentic article — must quite essentially include the sustained critical scrutiny of *ends*.

After the publication of *From Knowledge to Wisdom*, I turned for a time to grappling with the problems of quantum theory — partly, perhaps, in order to preserve my sanity. But then, in 1994, after early retirement from University College London (because of horrible things going on in my department) I turned again to the first crucial part of my argument concerning science, and wrote *The Comprehensibility of the Universe*. In this book I was able to solve a problem that had haunted me for two decades (a problem that defeated Einstein): What does it mean to assert of a theory that it is *unified*?[42] The book was published by Oxford University Press in 1998, received excellent reviews, but made no discernible impact on Philosophy of Science.

Since then, I have continued to develop and expound the "from knowledge to wisdom" argument in books, papers, and lectures. In 2003 I founded *Friends of Wisdom*, an international group of academics and educationalists, at the time of writing some 360 members strong, devoted to promoting wisdom in the university.[43] My own university now speaks of "the wisdom agenda" and "Developing a culture of wisdom at UCL" on its website[44] — the latter the title of a policy document which can be downloaded. There are many other signs that universities have recently begun to put some elements of wisdom-inquiry into academic practice.[45]

---

[41] Einstein (1973, p. 337).
[42] See Maxwell (1998, ch. 4). I also argued that aim-oriented empiricism solves the problems of induction and verisimilitude.
[43] See www.knowledgetowisdom.org/ (accessed 6 February 2014).
[44] See www.ucl.ac.uk/research/wisdom-agenda (accessed 6 February 2014).
[45] See note 10 on page 124.

Our long term failure to put wisdom-inquiry into practice is, as I see it, a monumental and very damaging *philosophical* blunder. For it is a blunder about what ought to be the aims and methods of inquiry, of learning. For too long we have unthinkingly taken for granted that inquiry ought to be, in the first instance, devoted to the pursuit of knowledge whereas, actually, it ought to have been devoted to helping us learn how to realize what is of value in life. Acquiring knowledge is important, but what humanity primarily needs to learn is how to live. Academic philosophers ought to be shouting from the rooftops about this profound philosophical blunder which, as we have seen, now threatens the future of humanity. At present, they are not. I urge my fellow philosophers to do what philosophy ought to do: devote reason to the task of helping to create a wiser world.

This, then, is my conclusion. Research in universities has been devoted, primarily, to acquiring knowledge and technological know-how. But these increase our power to act which, without wisdom, can lead to as much harm as benefit. Current global crises, and especially the most serious, global warming, have arisen in this way. We urgently need to bring about a revolution in our universities so that they come to seek and promote wisdom — wisdom being understood to be the capacity to realize what is of value in life, thus including knowledge, understanding, and technological know-how, but much else besides. Universities need to take up the task of helping humanity learn how to make progress towards as good a world as possible. There are signs that this revolution may already be underway. If so, it is happening with agonizing slowness, in a dreadfully muddled and piecemeal way. The underlying intellectual reasons for academic change need to be much more widely appreciated, to help give direction, coherence, and a rationale to this nascent academic revolution, and to help ensure that the intellectual value and integrity of science, scholarship, and education are strengthened and not subverted.

Acknowledgements

Permission to republish the following essays is gratefully acknowledged.

Chapter One: Philosophy Seminars for Five Year Olds, first published in *Learning for Democracy*, vol. 1, no. 2, 2005, pp. 71–7.

Chapter Two: What Philosophy Ought to Be, first published in C. Tandy, ed., 2014, *Death And Anti-Death, Volume 11: Ten Years After Donald Davidson (1917–2003)*, Ria University Press, Palo Alto, CA, ch. 7, pp. 125–62.

Chapter Three: How Can Our Human World Exist and Best Flourish Embedded in the Physical Universe? An Outline of a Problem-Based Liberal Studies Course, first published in an abbreviated version online in *On the Horizon*, vol. 22, issue 1, 2014, Emerald, special issue: Liberal Education in Crisis? Functions, Forms, and Foils. Guest Editor: Robert Bates Graber; www.emeraldinsight.com/journals.htm?issn=1074-8121&volume=22&issue=1.

Chapter Four: What's Wrong with Science and Technology Studies? What Needs to Be Done to Put It Right?, first published in Raffaelo Pisano, ed., 2014, *A Bridge between Conceptual Frameworks, Sciences, Society and Technology Studies*, Springer, Dordrecht.

Chapter Five: Arguing for Wisdom in the University: An Intellectual Autobiography, first published in *Philosophia*, vol. 40, no. 4, 2012, pp. 663–704.

References

Armstrong, D., 1978, *A Theory of Universals*, Cambridge University Press, Cambridge.
—, 1983, *What is a Law of Nature?*, Cambridge University Press, Cambridge.
Austin, J.L., 1962, *Sense and Sensibilia*, Oxford University Press, Oxford.
Ayer, A.J., 1960, *Language, Truth and Logic*, Gollancz, London (first published 1936).
Barnes, B., 1977, *Interests and the Growth of Knowledge*, Routledge & Kegan Paul, London.
—, 1982, *T.S. Kuhn and Social Science*, Macmillan, London.
—, 1985, *About Science*, Blackwell, Oxford.
Barnes, B., Bloor, D. and Henry, J., 1996, *Scientific Knowledge: A Sociological Analysis*, University of Chicago Press, Chicago.
Barnett, R. and Maxwell, N., eds., 2008, *Wisdom in the University*, Routledge, London.
Berkeley, G., 1957, *A New Theory of Vision and other writings*, Dent, London (first published 1709, 1710, and 1713).
Bloor, D., 1976, *Knowledge and Social Imagery*, Routledge and Kegan Paul, London.
Butterfield, H., 1951, *The Whig Interpretation of History*, Bell and Sons, London (first published in 1931).
Chalmers, D., 1996, *The Conscious Mind*, Oxford University Press, Oxford.
Crossman, R., 1937, *Plato Today*, George Allen and Unwin, London.
Dennett, D., 1991, *Consciousness Explained*, Allen Lane, London.
Descartes. R., 1949, *A Discourse on Method Etc.*, Dent, London (first published 1637).

Dretske, F., 1977, Laws of Nature, *Philosophy of Science* 44, pp. 248–68.

Eddington, A., 1947, *The Nature of the Physical World*, Dent, London.

Einstein, A., 1949, Autobiographical Notes, in Schilpp, P.A., ed., *Albert Einstein: Philosopher-Scientist*, Open Court, La Salle.

—, 1973, *Ideas and Opinions*, Souvenir Press, London.

Feyerabend, P., 1965, Problems of Empiricism, in Colodny, R.G., ed., *Beyond the Edge of Certainty*, Prentice-Hall, New York, pp. 145–260.

—, 1970, Problems of Empiricism, Part II, in Colodny, R.G., *The Nature and Function of Scientific Theories*, University of Pittsburgh Press, Pittsburgh, pp. 275–354.

—, 1975, *Against Method: Outline of an Anarchistic Theory of Knowledge*, New Left Books, London.

—, 1978, *Science in a Free Society*, New Left Books, London.

—, 1987, *Farewell to Reason*, Verso, London.

Gay, P., 1973, *The Enlightenment: An Interpretation*, Wildwood House, London.

Gross, P. and Levitt, N., 1994, *Higher Superstition: The Academic Left and Its Quarrels with Science*, John Hopkins University Press, Baltimore.

Gross, P., Levitt, N. and Lewis, M., eds., 1996, *The Flight from Science and Reason*, John Hopkins University Press, Baltimore.

Guthrie, W.K.C., 1978, *A History of Greek Philosophy: Vol. II, The Presocratic Tradition from Parmenides to Democritus*, Cambridge University Press, Cambridge.

Hardy, A., 1965, *The Living Stream*, Collins, London.

Harper, W.L., 2011, *Isaac Newton's Scientific Method*, Oxford University Press, Oxford.

Hawking, S. and Mlodinow, L., 2010, *The Grand Design*, Bantam Books, London.

Howson, C., 2000, *Hume's Problem*, Oxford University Press, Oxford.

Hume, D., 1959, *A Treatise of Human Nature*, Book 1, Everyman, London (first published 1738).

Jackson, F., 1982, Epiphenomenal Qualia, *Philosophical Quarterly* 32, pp. 127–36.

—, 1986, What Mary Didn't Know, *Journal of Philosophy 3*, pp. 291–95.

Kant, I., 1950, *Critique of Pure Reason*, Dent, London (second edition first published 1787).

—, 1953, *Prolegomena to any Future Metaphysics that will be able to present itself as a Science*, Manchester University Press, Manchester (first published 1783).

Koertge, N., 1998, *A House Built on Sand: Exposing Postmodernist Myths About Science*, Oxford University Press, Oxford.

Kuhn, T.S., 1962, *The Structure of Scientific Revolutions*, Chicago University Press, Chicago.

Kyburg, H., 1970, *Probability and Inductive Logic*, Collier-Macmillan, Toronto.

Lakatos, I., 1970, Falsification and the Methodology of Scientific Research Programmes, in Lakatos, I. and Musgrave, A., eds., *Criticism and the Growth of Knowledge*, Cambridge University Press, London, pp. 91–195.

Lakatos, I. and Musgrave, A., eds., 1970, *Criticism and the Growth of Knowledge*, Cambridge University Press, London.

Locke, J., 1961, *An Essay Concerning Human Understanding*, Dent, London (first published 1690).

Longuet-Higgins, C., 1984, For Goodness Sake, *Nature 312*, p. 204, available online at www.ucl.ac.uk/from-knowledge-to-wisdom/reviews/#goodness.

Magee, B., 1997, *Confessions of a Philosopher*, Weidenfeld and Nicolson, London.

Matthews, G., 1980, *Philosophy for the Young Child*, Harvard University Press, Harvard.

Matthews, M.R., 1989, *The Scientific Background to Modern Philosophy*, Hackett, Indianapolis.

Maudlin, T., 2010, The Geometry of Space-Time, *The Aristotelian Society, Supplementary vol. LXXXIV*, pp. 63–78.

Maxwell, N., 1966, Physics and Common Sense, *British Journal for the Philosophy of Science 16*, pp. 295–311.

—, 1968a, Can there be Necessary Connections between Successive Events?, *British Journal for the Philosophy of Science 19*, pp. 1–25.

—, 1968b, Understanding Sensations, *Australasian Journal of Philosophy 46*, pp. 127–46.

—, 1971, Clash of Ideas, review of I. Lakatos and A. Musgrave, eds., *Criticism and the Growth of Knowledge*, Cambridge University Press, Cambridge, 1970, in *Nature 231*, p. 269.

—, 1972a, A Critique of Popper's Views on Scientific Method, *Philosophy of Science 39*, pp. 131–52.

—, 1972b, A New Look at the Quantum Mechanical Problem of Measurement, *American Journal of Physics 40*, pp. 1431–5.

—, 1973a, Alpha Particle Emission and the Orthodox Interpretation of Quantum Mechanics, *Physics Letters 43A*, pp. 29–30.

—, 1973b, The Problem of Measurement—Real or Imaginary?, *American Journal of Physics 41*, pp. 1022–5.

—, 1974, The Rationality of Scientific Discovery, *Philosophy of Science 41*, pp. 123–53 and 247–95.

—, 1976a, *What's Wrong with Science?*, Bran's Head Books, Frome.

—, 1976b, Towards a Micro Realistic Version of Quantum Mechanics, *Foundations of Physics 6*, pp. 275–92 and 661–76.

—, 1980, Science, Reason, Knowledge and Wisdom: A Critique of Specialism, *Inquiry 23*, no. 1, pp. 19–81.

—, 1982, Instead of Particles and Fields, *Foundations of Physics 12*, pp. 607–31.

—, 1984, *From Knowledge to Wisdom*, Basil Blackwell, Oxford.

—, 1988, Quantum Propensiton Theory: A Testable Resolution of the Wave/Particle Dilemma, *British Journal for the Philosophy of Science 39*, pp. 1–50.

—, 1992, What Kind of Inquiry Can Best Help Us Create a Good World?, *Science, Technology and Human Values 17*, pp. 205–27.

—, 1993a, Induction and Scientific Realism, *British Journal for the Philosophy of Science 44*, pp. 61–79, 81–101 and 275–305.

—, 1993b, Beyond Fapp: Three Approaches to Improving Orthodox Quantum Theory and An Experimental Test, in van der Merwe, A., Selleri, F. and Tarozzi, G., eds., *Bell's Theorem and the Foundations of Modern Physics*, World Scientific, pp. 362–70.

—, 1994, Particle Creation as the Quantum Condition for Probabilistic Events to Occur, *Physics Letters A 187*, pp. 351–5.

—, 1995, A Philosopher Struggles to Understand Quantum Theory: Particle Creation and Wavepacket Reduction, in Ferrero, M. and van der Merwe, A., eds., *Fundamental Problems in Quantum Physics*, Kluwer Academic, London, pp. 205-14.

—, 1998, *The Comprehensibility of the Universe: A New Conception of Science*, Oxford University Press, Oxford.

—, 1999, Are there Objective Values?, *The Dalhousie Review 79*, no. 3, pp. 301-17.

—, 2000a, Can Humanity Learn to Become Civilized? The Crisis of Science without Civilization, *Journal of Applied Philosophy 17*, no. 1, pp. 29-44.

—, 2000b, The Mind-Body Problem and Explanatory Dualism, *Philosophy 75*, pp. 49-71.

—, 2000c, A new conception of science, *Physics World 13*, no. 8, pp. 17-8.

—, 2001, *The Human World in the Physical Universe: Consciousness, Free Will and Evolution*, Rowman and Littlefield, Lanham.

—, 2002, The Need for a Revolution in the Philosophy of Science, *Journal for General Philosophy of Science 33*, pp. 381-408.

—, 2003, Do Philosophers Love Wisdom, *The Philosophers' Magazine*, Issue 22, 2nd quarter, pp. 22-4.

—, 2004a, *Is Science Neurotic?*, Imperial College Press, London.

—, 2004b, Does Probabilism Solve the Great Quantum Mystery?, *Theoria 19/3*, no. 51, pp. 321-36.

—, 2005, Popper, Kuhn, Lakatos and Aim-Oriented Empiricism, *Philosophia 32*, nos. 1-4, pp. 181-239.

—, 2006, Practical Certainty and Cosmological Conjectures, in Rahnfeld, M., ed., *Gibt es sicheres Wissen?*, Leipziger Universitätsverlag, Leibzig, pp. 44-59.

—, 2007a, *From Knowledge to Wisdom: A Revolution for Science and the Humanities*, Pentire Press, London (2nd revised and extended edition of Maxwell, 1984).

—, 2007b, From Knowledge to Wisdom: The Need for an Academic Revolution, *London Review of Education 5*, pp. 97-115 (reprinted in Barnett and Maxwell, 2008, pp. 1-19).

—, 2008a, Do We Need a Scientific Revolution?, *Journal for Biological Physics and Chemistry 8*, no. 3, pp. 95-105.

—, 2008b, Are Philosophers Responsible for Global Warming?, *Philosophy Now* 65, pp. 12–3.

—, 2009a, How Can Life of Value Best Flourish in the Real World?, in McHenry, L., ed., *Science and the Pursuit of Wisdom: Studies in the Philosophy of Nicholas Maxwell*, Ontos Verlag, Frankfurt, pp. 1–56.

—, 2009b, Are Universities Undergoing an Intellectual Revolution?, *Oxford Magazine 290*, Eighth Week, Trinity Term, June, pp. 13–6.

—, 2009c, The Metaphysics of Science: An Account of Modern Science in Terms of Principles, Laws and Theories (review of book by Craig Dilworth), *International Studies in the Philosophy of Science 23*, no. 2, pp. 228–32.

—, 2009d, *What's Wrong With Science?*, 2nd ed., Pentire Press, London.

—, 2010a, *Cutting God in Half – And Putting the Pieces Together Again: A New Approach to Philosophy*, Pentire Press, London (available free online at http://discovery.ucl.ac.uk/view/people/ANMAX22.date.html).

—, 2010b, Reply to Comments on *Science and the Pursuit of Wisdom*, *Philosophia 38*, no. 4, pp. 667–90.

—, 2011a, Three Philosophical Problems about Consciousness and their Possible Resolution, *Open Journal of Philosophy 1*, no. 1, pp. 1–10.

—, 2011b, A Priori Conjectural Knowledge in Physics, in Schaffer, M. and Veber, M., eds., *What Place for the A Priori?*, Open Court, Chicago, pp. 211–40.

—, 2011c, Is the Quantum World Composed of Propensitons?, in Suárez, M., ed., *Probabilities, Causes and Propensities in Physics*, Springer, Dordrecht, pp. 219–41.

—, 2012a, In Praise of Natural Philosophy: A Revolution for Thought and Life, *Philosophia 40*, no. 4, pp. 705–15.

—, 2012b, How Universities Can Help Humanity Learn How to Resolve the Crises of Our Times – From Knowledge to Wisdom: The University College London Experience, *Handbook on the Knowledge Economy 2*, Heam, G., Katlelle, T. and Rooney, D., eds., Edward Elgar, Cheltenham, pp. 158–79.

—, 2013a, Has Science Established that the Cosmos is Physically Comprehensible?, in Travena, A. and Soen, B., eds., *Recent Advances in Cosmology*, Nova Science Publishers Inc., New York.

—, 2013b, Does Philosophy Betray Both Reason and Humanity? (renamed without permission "Knowledge or Wisdom"), *Philosophers' Magazine 62*, pp. 17–8.

—, 2014a, *How Universities Can Help Create a Wiser World: The Urgent Need for an Academic Revolution*, Imprint Academic, Exeter.

—, 2014b, Unification and Revolution: A Paradigm for Paradigms, *Journal for General Philosophy of Science 45*, issue 1, pp. 133–49, available online at http://philpapers.org/rec/MAXUAR.

—, forthcoming, Popper's Paradoxical Pursuit of Natural Philosophy, in Shearmur, J. and Stokes, G., eds., *Cambridge Companion to Popper*, Cambridge University Press, Cambridge.

McHenry, L., ed., 2009a, *Science and the Pursuit of Wisdom: Studies in the Philosophy of Nicholas Maxwell*, Ontos Verlag, Frankfurt.

—, 2009b, Ghosts in the Machine: Comment of Sismondo, *Social Studies of Science 39*, pp. 943–7.

Midgley, M., 1986, Is Wisdom Forgotten?, *University Quarterly: Culture, Education and Society 40*, pp. 425–7.

Moore, G., 1959, *Philosophical Papers*, Allen and Unwin, London.

Nagel, T., 1974, What is it Like to Be a Bat?, *Philosophical Review 83*, pp. 435–50.

—, 1989, *The View from Nowhere*, Oxford University Press, Oxford.

Newton, I., 1962, *Principia*, A. Motte's translation revised by F. Cajori, University of California Press, Berkeley (first published in 1687).

Peierls, R.E. and Enogat, J., eds., 1947, *Science News 2*, Penguin Books, Harmondsworth.

Popper, K., 1959, *The Logic of Scientific Discovery*, London: Hutchinson.

—, 1962, *The Poverty of Historicism*, Routledge and Kegan Paul, London.

—, 1963, *Conjectures and Refutations*, Routledge and Kegan Paul, London.

—, 1969, *The Open Society and Its Enemies*, Routledge and Kegan Paul, London.

—, 1970, Normal Science and Its Dangers, in Lakatos, I. and Musgrave, A., eds., *Criticism and the Growth of Knowledge*, Cambridge University Press, London, pp. 51–8.
—, 1976, *Unended Quest*, Fontana, London.
Russell, B., 1905, On Denoting, *Mind*, new series, 14, pp. 479–93.
—, 1925, *The ABC of Relativity*, Kegan Paul, London.
—, 1946, *History of Western Philosophy*, George Allen and Unwin, London.
—, 1956, *Logic and Knowledge*, Marsh, R.C., ed., Allen and Unwin, London.
Ryle, G., 1949, *The Concept of Mind*, Hutchinson, London.
Sinclair, W.A., 1945, *An Introduction to Philosophy*, Oxford University Press, Oxford.
Singer, P., 1995, *Animal Liberation*, Pimlico, London.
Sismondo, S., 2009a, Ghosts in the Machine: Publication Planning in the Medical Sciences, *Social Studies of Science 39*, pp. 171–98.
—, 2009b, Ghosts in the Machine: Response to McHenry, *Social Studies of Science 39*, pp. 949–52.
Smart, J.J.C., 1963, *Philosophy and Scientific Realism*, Routledge and Kegan Paul, London.
Snow, C.P., 1964, *The Two Cultures and a Second Look*, Cambridge University Press, Cambridge.
Sokal, A., 1998, Transgressing the Boundaries; Toward a Transformative Hermeneutics of Quantum Gravity, in Sokal, A. and Bricmont, B., *Intellectual Impostures*, Profile Books, London, pp. 199–240.
—, 2008, *Beyond the Hoax: Science, Philosophy and Culture*, Oxford University Press, New York.
Sokal, A. and Bricmont, B., 1998, *Intellectual Impostures*, Profile Books, London.
Swain, M., ed., 1970, *Induction, Acceptance and Rational Belief*, Reidel, Dordrecht.
Tooley, M., 1977, The Nature of Law, *Canadian Journal of Philosophy 7*, pp. 667–98.
van Fraassen, B., 1989, *Laws and Symmetry*, Clarendon Press, Oxford.
Voltaire, 1980, *Letters on England*, Penguin, Harmondsworth (first published in English in 1733, and in French in 1734).

Weinberg, S., 1993, *Dreams of a Final Theory*, Hutchinson, London.
Whitehead, A.N., 1932, *Science and the Modern World*, Cambridge University Press, Cambridge.
Wittgenstein, L., 1958, *Philosophical Investigations*, Blackwell, Oxford.
—, 1960, *Tractatus Logico-Philosophicus*, Routledge and Kegan Paul, London (first published 1922).
Ziman, J., 1968, *Public Knowledge*, Cambridge University Press, Cambridge.

Index

academic inquiry 2–3, 98–107
  and literature 3
  and problem solving 2–3,
    49–51
  and problems of living 7–
    8, 19–22, 100–3, 109
  and specialized
    knowledge 6–7, 54–5, 101
  irrationality of 15–18, 45,
    98–101, 117
academic revolution viii, 8,
  17–22, 40, 46, 61, 98–107,
  109, 118–24, 171–5
aim-oriented empiricism 28–
  31, 92–7, 104, 109, 120–1,
  164–9
aim-oriented rationality 61–
  4, 104–5, 109, 121–4, 168–
  72, 174–5
aims, problematic character
  of 29, 61–4, 65, 88–9, 93–7,
  104–5, 120–3, 138–41, 145–
  8, 165–6, 168–72, 174–6
Alembert, J. d' 118
Anaximander 23
anthropology 54, 115
Aristotle 23, 78
Armstrong, D. 159–60

Austen, J. 132
Austin, J.L. 44n, 150
Ayer, A.J. 136

Bacon, F. 115
Balzac, H. de 132
Barnes, B. 74
Beckett, S. 133
Bergman, I. 133
Berkeley, G. 33–4, 41, 44, 58n
Berlin, I. 117
Bethe, H. 130
Bloor, D. 74
Bondi, H. 153
Boyle, R. 23
Bradley, F.H. 35
Bricmont, J. 85
Bronte, E. 132–3
Butterfield, H. 79
Bynum, W. 83–4

Carnap, R. 36, 72
Cartesian dualism 32–3, 35,
  41, 42, 44, 58, 141–2, 158
causation 21–2, 155–6
Chalmers, D. 40
Chekhov, A. 132

childhood
    learning 1–2, 49–51, 128–9
    problems 125–9
climate change viii, 18, 21,
    45, 61, 112, 174, 176
Collins, H. 74n
Condorcet, N. de 103, 114
Conrad, J. 132
Crossman, R. 146n

Darwin, C. 42, 52, 78
Deleuze, G. 85
Democritus 23, 57
Dennett, D. 40
Derrida, J. 78
Descartes, R. 23, 24, 25, 26,
    31, 32–3, 40, 41, 44, 57–8
desires, riddle of 137–47, 171
Diderot, D. 103, 114
Dostoevsky, F. 132–3
Dretske, F. 159–60

Eccles, J. 153
economics 14n, 54, 69, 168
Eddington, A. 131, 132, 134
education vii, 47–64, 129
    and problem solving 2–3,
    7–8, 49–51, 135–6
Einstein, A. ix, 50, 52, 61, 166,
    175
Eliot, G. 132
Enlightenment Programme
    114–24, 169
    critics of 117
    failures of 116–22, 169–70
    new 104–5, 120–4, 170–2
    three steps of 104–5, 116
    traditional 115–7

Enlightenment, The 103–5,
    109, 114
evolution 5, 42, 44, 47, 82n,
    168

Faraday, M. 52, 73n, 78
Feigl, H. 36
Feyerabend, P. 69, 72–4, 162
Fichte, J. 35
Fielding, H. 132
Fitzgerald, F. Scott 132
Foucault, M. 78
Frank, P. 36
Freud, S. 134–5
Frisch, O.R. 130
fundamental problem vii,
    13–15, 26, 31–5, 37, 39, 41,
    47–9, 55–7, 155–9
    and Darwin 42
    and specialized research
    54–5
    experiences in response to
    137–45
    outline of proposed
    solution 42–4, 59–60, 155–9
    proposed solutions 57–60
    see also human world/
    physical universe problem

Galileo 23, 25, 26, 31, 32, 57–8
Gay, P. 169
general relativity 53–4
global problems vii–viii, 11–
    12, 17–21, 40, 51, 60–2, 99,
    101, 107, 174
    role of science 21–2, 60–2,
    112
God, desire to be 138–48

Grand Challenges
  Programme at UCL 16n, 175
Green, T.H. 35
Gross, P. 85
Grünbaum, A. 135n

habitat destruction vii, 18, 61, 174
Hardy, A. 82
Hardy, T. 132
Harper, 71n
Hawking, S. 27
Hegel, G. 35
Heidegger, M. 35
Hempel, C. 36, 72
Henry, J. 74n
Heraclitus 23
history and philosophy of science 69–77
  decline of 70–89
history of progress 78–84
Hobbes, T. 23, 26
Hoyle, F. 132
human world 132–5
  and literature 132–4
human world/physical universe problem 31–5, 37, 39, 41, 47–9, 55–7, 155–9
  and Darwin 42
  and specialized research 54–5
  outline of proposed solution to 42–4, 59–60, 155–9
  proposed solutions 57–60
  see also fundamental problem

humanities 3, 14, 48–9, 51, 98, 102, 169–71
Hume, D. 27, 34, 41, 44, 131, 137, 155
  refutation of view of causation 155–6, 158–60
Husserl, E. 35
Huygens, C. 23, 26, 31, 57

Ibsen, H. 132
idealism 35, 58–9
Iliffe, R. 84
induction, problem of 26–31, 74, 131–2, 136–7
  solution to 28–31, 94–5
inequality vii, 21, 61, 112, 174

Jackson, F. 43n, 56n, 159
Jeans, J. 132
Joyce, J. 132

Kafka, F. 132–3
Kant, I. 23, 27, 34–5, 41, 44
Keats, J. 144
Kepler, J. 23, 25, 26
knowledge-inquiry 20, 98–103, 111–2
  damaging irrationality of viii, 7, 20–2, 98–101, 174–5
  mixed benefits of 21, 111–2
Koertge, N. 85–6
Kuhn, T.S. 70, 72, 73n, 74, 162

Lacan, J. 85
Lakatos, I. 70, 162
Latour, B. 85
Laudan, L. 69
Lawrence, C. 83–4
Lawrence, D.H. 132
Leibniz, G. 23, 26

Levitt, N. 85
Locke, J. 23, 26, 33, 41, 44, 57
logical atomism 35–6
logical empiricism 72–3
logical positivism 36–8, 72n
Longuet-Higgins, C. 28n, 106–7, 173–4

Magee, B. 128
Mann, T. 132
Marx, K. 68, 122, 153
Matthew, G. 9
Maudlin, T. 40
Mauriac, F. 132
McHenry, L. 87n
McTaggart. J. 35
Medawar, P. 153
Merleau-Ponty, M. 35
metaphysics 35–7, 63, 91–3, 95
Midgley, M. 173
Mill, J.S. 122
mind/body problem 32–3, 44, 60, 136, 158
Moore, G.E. 35, 37–8, 136
mystical experience 143–5

Nagel, E. 72
Nagel, T. 40, 43n, 56n, 159
national curriculum of UK 9–10
natural philosophy 23–6, 120
   and aim-oriented empiricism 30–1, 120
   demise of 25–6
Neurath, O. 36
Neve, M. 83–4

Newton, I. 23–6, 30, 41n, 57, 78
   Principia 23–5

Orwell, G. 132

Peierls, R. 130
philosophy 135–7
   analytic 36, 38–9, 41, 64, 149–50
   and children 4–6, 9
   and fundamental problem vii, 64
   and science 23–31, 32
   basic task of 11, 12–17
   central role in education 3–6, 64
   continental 38–9, 41, 64
   empirical 123
   failures of 11, 16–17, 20, 26, 33–41, 44–5, 64, 176
   megalomaniac impulses of 145–8
philosophy of science 67–97, 105
   and specialization 84–5
physical universe 130–2
physics 42–4
   and experiential world 42–4, 156–60
   and metaphysics 63, 91–3, 95, 119–20
   and necessitating properties 155–6, 159
   and theoretical unity 43, 62–3, 89–92, 119, 162–6
   basic task of 42–4
   childhood interest in 130–2
   incompleteness of 42–4,

152, 155–9
see also aim-oriented empiricism, standard empiricism
Plato 23, 41, 68, 111, 146–7, 153
play, role in learning viii–ix
pollution vii, 18, 61, 112
Popper, K. 28, 39–40, 53n, 67–70, 74, 118, 135n, 170
 and critical rationalism 68, 109, 152, 167–8
 and falsificationism 67–8, 109, 162, 167–8
 and the open society 68, 146n, 153
 and theoretical unity 162–4
 and truth 108, 150–2
 criticisms of 162–4
 significance of 150–4
population growth vii, 61, 112
propositional calculus 135

quantum theory 161, 175

reason 99
 and wisdom 117
 rules of 15–16, 40, 99–101, 170
 see also aim-oriented rationality
Reichenbach, H. 36
relativism 148
Russell, B. 35–8, 40, 131, 136, 146n
Ryle, G. 39, 44n, 149

Sartre, J.-P. 35

science
 aims of 62–3, 71, 94–7, 165–7
 and metaphysics 63, 89–94, 164–5
 and philosophy 23–31, 95–7
 and politics 63, 97–8, 166–7
 social character of 75–7
 and values 63, 76–7, 95–8, 120, 166–7
 and wisdom 111–4
science and technology studies 65–6, 84, 86–9
 proper task of 106–7
science wars 71, 85–7
scientific progress 67–9, 74, 78–84, 103, 167
Schelling, F. 35
Schleiermacher, F. 35
Schlick, M. 36
Schopenhauer, A. 35
self, discovery of 138–45
Shakespeare, W. 133
Shaw. G. 132
Sinclair, W.A. 131
Singer, P. 40
Sismondo, S. 86–7
Smart, J.J.C. 40, 58n, 154–5
Snow, C.P. 124
social constructivism 77–9
 criticism of 78–87
social inquiry 14, 48–9, 51, 101–3, 105, 122, 170–1
social methodology 105, 122–3, 169

social progress 89, 103–4, 114–6, 122, 167
social science 41n, 98, 101–5, 115–6, 122, 169–70
sociology 105, 115
sociology of science 71, 74–87, 95, 105
Socrates 23, 111, 146
Sokal, A. 28n, 65, 85
specialization, evils of 3–4, 17–18, 84–5
species, extinction of vii, 18, 61, 112, 174
Spinoza, B. 23, 26
standard empiricism 87–9, 104–5, 118
  refutation of 89–92, 118–9, 164–5
Stendhal 132
Stevenson, R.L. 133
Strindberg, A. 133
strong programme 74–7
symbolic logic 135

Teller, E. 130
Thales 23
Tolstoy, L. 132
Tooley, M. 159–60
Turgenev, I. 132

van Fraassen, B. 160
Voltaire 24, 103, 114

Waismann, F. 36
war viii, 5, 7, 18, 21–2, 45, 61, 80, 112, 125, 174
Weinberg, S. 27
Wells, H.G. 133
Whiggish history 79–84
  bad consequences of 83–4
  how to oppose 79–83
Whitehead, A.N. 41
wisdom viii, 103, 118, 173
  aim of inquiry viii, 8, 103–5
  aim of philosophy 111
  and reason 117
  and science 111, 113–4
  definition of viii, 103, 118
  friends of 175
  key to 110–1, 113–4, 115n
  need for 112–3
wisdom-inquiry 18–20, 101–7, 113–5, 117, 122–4, 170–6
  synthesis of rationalism and romanticism 123–4
  see also aim-oriented empiricism, aim-oriented rationalism
Wittgenstein, L. 35, 36, 38
Woolf, V. 132–3

Ziman, J. 27

What critics have said about two of the author's previous books:

From Knowledge to Wisdom
Maxwell is advocating nothing less than a revolution (based on reason, not on religious or Marxist doctrine) in our intellectual goals and methods of inquiry… There are altogether too many symptoms of malaise in our science-based society for Nicholas Maxwell's diagnosis to be ignored.
Professor Christopher Longuet-Higgins, *Nature.*

The essential idea is really so simple, so transparently right… It is a profound book, refreshingly unpretentious, and deserves to be read, refined and implemented.
Dr. Stewart Richards, *Annals of Science.*

…a strong effort is needed if one is to stand back and clearly state the objections to the whole enormous tangle of misconceptions which surround the notion of science today. Maxwell has made that effort in this powerful, profound and important book.
Dr. Mary Midgley, *University Quarterly.*

This book is a provocative and sustained argument for a 'revolution', a call for a 'sweeping, holistic change in the overall aims and methods of institutionalized inquiry and education, from knowledge to wisdom'… Maxwell offers solid and convincing arguments for the exciting and important thesis that rational research and debate among professionals concerning values and their realization is both possible and ought to be undertaken.
Professor Jeff Foss, *Canadian Philosophical Review.*

Wisdom, as Maxwell's own experience shows, has been outlawed from the western academic and intellectual system… In such a climate, Maxwell's effort to get a hearing on behalf of wisdom is indeed praiseworthy.
Dr. Ziauddin Sardar, *Inquiry.*

Maxwell has, I believe, written a very important book which will resonate in the years to come. For those who are not inextricably and cynically locked into the power and career structure of academia with its government-industrial-military connections, this is a book to read, think about, and act on.
Dr. Brian Easlea, *Journal of Applied Philosophy.*

Maxwell's argument… is a powerful one. His critique of… the philosophy of knowledge is coherent and well argued, as is his defence of the philosophy

of wisdom... This is an exciting and important work.
Dr. John Hendry, *British Journal for the History of Science.*

This book is written in simple straightforward language... The style is passionate, committed, serious; it communicates Maxwell's conviction that we are in deep trouble, that there is a remedy available, and that it is ingrained bad intellectual habits that prevent us from improving our lot... Maxwell is raising an important and fundamental question and things are not going so well for us that we should afford the luxury of listening only to well-tempered answers.
Professor John Kekes, *Inquiry.*

Any philosopher or other person who seeks wisdom should read this book. Any educator who loves education — especially those in leadership positions — should read this book. Anyone who wants to understand an important source of modern human malaise should read this book. And anyone trying to figure out why, in a world that produces so many technical wonders, there is such an immense 'wisdom gap' should read this book... Maxwell presents a compelling, wise, humane, and timely argument for a shift in our fundamental 'aim of inquiry' from that of knowledge to that of wisdom.
Jeff Huggins, *Metapsychology.*

Is Science Neurotic?

Maxwell has written a very important book... [He] eloquently discusses the astonishing advances and the terrifying realities of science without global wisdom. While science has brought forth significant advancements for society, it has also unleashed the potential for annihilation. Wisdom is now, as he puts it, not a luxury but a necessity.
Professor Joseph Davidow, *Learning for Democracy.*

Is science neurotic? Yes, says Nicholas Maxwell, and the sooner we acknowledge it and understand the reasons why, the better it will be for academic inquiry generally and, indeed, for the whole of humankind. This is a bold claim... But it is also realistic and deserves to be taken very seriously... I found the book fascinating, stimulating and convincing... I have come to see the profound importance of its central message.
Dr. Mathew Iredale, *The Philosopher's Magazine.*